SCIENCE AS A CULTURAL PROCESS

SCIENCE AS A CULTURAL PROCESS

MAURICE N. RICHTER, JR.

SCHENKMAN PUBLISHING COMPANY, INC.
Cambridge, Massachusetts
Distributed by General Learning Press

Schenkman books are distributed by
GENERAL LEARNING PRESS
250 James Street
Morristown, New Jersey 07960

Library of Congress Catalog Card Number: 75—182188

Copyright © 1972
SCHENKMAN PUBLISHING COMPANY, INC.
Cambridge, Massachusetts 02138

Printed in the United States of America

All rights reserved. This book, or parts thereof, may not be reproduced in any form without written permission of the publishers.

To my parents

CONTENTS

ACKNOWLEDGMENTS vii

INTRODUCTION ix

CHAPTER ONE: AN APPROACH TO THE STUDY OF SCIENCE 1

CHAPTER TWO: THE CONCEPT OF SCIENCE IN SOCIOLOGICAL PERSPECTIVE 9
The Scientist, the Sociologist, and the Sociology of Science 9
The Conception of Science as a Method 14
The Conception of Science as a Social Institution 17
The Conception of Science as an Occupation 20
The Conception of Science as a Profession 25
An Alternative Conception of Science 33

CHAPTER THREE: SCIENCE AS A CULTURAL PROCESS 35
Science as Cultural Counterpart of Individual Cognitive Development 36
Science as an Outgrowth of Traditional Cultural Knowledge 43
Science as a Cognitive Form of Cultural Development 53
Summary 57

CHAPTER FOUR: ASPECTS OF THE DEVELOPMENT OF SCIENCE — 62
"Science" and "Development of Science" — 62
Problems in the Study of the Development of Science — 63
A Preliminary Framework — 66
The Cultural Background: Rationalism and Empiricism — 67
The Emergence of Scientific Systems — 71
The Emergence of the Scientific Process — 77

CHAPTER FIVE: THE CHANGING RELATIONSHIP BETWEEN SCIENCE AND SOCIETY — 86
Requirements for the Persistence of Science — 86
The Growth and Impact of Science — 90
Possible Futures of Science — 95

CHAPTER SIX: THE ORGANIZATION OF THE SCIENTIFIC COMMUNITY — 100
The Concept of "Scientific Community" — 100
The Scientific Community and External Structures — 103
The Scientific Exchange System — 105
The Normative System of Science — 108
Some Complexities in the Scientific Normative System — 111
Some Organizational Implications — 116

CONCLUSION — 125

ACKNOWLEDGMENTS

My interest in the sociology of science was stimulated by three of my teachers, Gerard DeGré, Anselm Strauss, and Edward Shils. Several of my colleagues at the State University of New York at Albany have given me helpful suggestions: these include Harold F. Blum, Mason Griff, Alicja Iwanska, Stuart H. Palmer, Freddie O. Sabghir, and Paul F. Wheeler. I am also grateful for detailed comments on the manuscript, from Ralph A. W. Lehman, and from my brothers, Marcel K. Richter and Wayne H. Richter. I have benefited from discussions on various topics mentioned in the book, with Richard C. Stillman and Robert Lamson. I am, of course, exclusively responsible for all errors and shortcomings in the book.

I am grateful for facilities made available by the State University of New York at Albany, and in particular for the assistance of Paul Meadows, Chairman of the Department of Sociology, and Paul F. Wheeler, Associate Dean of the College of Arts and Sciences. I have also received considerable support and encouragement from my parents, Maurice N. Richter and Brina Kessel Richter.

INTRODUCTION

This book is an attempt by a sociologist to analyze science as a social phenomenon. Attention is to be directed not to particular branches of science, or to science at particular times or places, but rather to science as a whole, as a process extending across national boundary lines and across centuries, and manifesting itself in numerous disciplines and in a variety of social contexts. The purpose is not to provide new facts about science, or to propose or test any sociological hypothesis, but rather to contribute to the clarification of the sociological meaning of science: to relate the concept of science to the conceptual framework of sociology, and to identify ways in which science is similar to, and different from, various other phenomena which sociologists study. Features of science which mark it as unique among social phenomena, and which render various familiar sociological concepts only imperfectly applicable to it, make this a challenging task.

The questions to be discussed here are broad enough in their significance to make it reasonable to hope that the book will have some interest not only for sociologists of science, but also for scientists whose general interest in sociology is minimal, and for sociologists specializing in the study of phenomena far removed from science, as well as for members of the wider public who are concerned with the critical problems involved in the relation between science and society.

I shall assume that the reader has some familiarity with science, and with sociological concepts. However, readers who do not have these qualifications should nevertheless be able to understand much of what is said here.

Chapter 1

AN APPROACH TO THE STUDY OF SCIENCE

I shall begin by suggesting a preliminary definition of science broad and vague enough to avoid pre-judging the issues to be discussed in the following pages, but, hopefully, narrow and clear enough to indicate the approximate boundaries of our subject matter, and reasonably consistent with familiar usage. Science is to be provisionally defined here as the process, or the group of inter-related processes, through which we have acquired our modern, and ever-changing, knowledge of the natural world which encompasses inanimate nature, life, human nature and human society. Knowledge acquired through this process may be called "scientific," with the understanding that knowledge accepted as scientific at one time is likely to be obsolete at a later date. Those who participate in science in relatively direct and creative ways may be called *scientists*. All scientists, insofar as they communicate openly among themselves about their respective scientific activities, may be recognized as participants in the *scientific community*.

This definition of science cautiously avoids commitments on some important issues related to the concept of science. It does not even specifically identify the process to be called "science," but says only that that term is to be applied to whatever process may be recognized as responsible for our modern knowledge of nature. It is thus broad enough to permit, within its framework, a variety of alternative conceptions of science.

On the other hand, this definition *does* involve a strong commitment in one sociologically important respect. It commits us to a conception of science based on standards which prevail in our own time and our own culture, even though science, as thus defined, has existed to some extent in times and cultures other than our own. It means, for example, that if we speak of science in ancient Greece, we are referring not to whatever process the Greeks considered appropriate for investigation of nature, but rather to whatever activities in ancient Greek society appear to us, retrospectively, as having involved movement toward our own modern knowledge of nature, though these activities might not have been clearly distinguished from other activities by the Greeks themselves.

Science as thus defined does have historical roots in several ancient cultures, especially that of Greece. However, it first emerged in what is known as its *modern* form in western Europe in the seventeenth century. The transition from pre-modern to modern science involved several inter-related aspects. Science came to be associated with a distinctive view of nature as operating according to general laws which remain largely hidden under ordinary observational circumstances but which can be uncovered through systematically controlled observation and experimentation. One set of proposed laws, formulated by Sir Isaac Newton, provided an overwhelmingly impressive demonstration of the potentialities of this approach. Science became reasonably clearly differentiated from such related phenomena as philosophy, religion, technology and magic. The role of "scientist" became differentiated from other roles, and scientists came to communicate with each other systematically in ways which mark the beginning of what we now call the "scientific community." After many centuries of irregular growth and decline, science came to acquire a self-reinforcing capacity which has been reflected in the massive, sustained growth of science from the seventeenth century to our own time.

These various inter-related developments, coming at about the same time as aspects of a single pattern, constituted such a profound transformation that for some purposes it is most convenient to define science in a way which excludes what came earlier. I have proposed here a definition which is somewhat more inclusive in its implications, but which must be used with the understanding that the science of centuries before the seventeenth was only very imperfectly distinguished from other phenomena which we today would not regard as science.

From its western European birthplace, modern science has spread widely around the world, although it remains heavily concentrated in regions of European cultural ancestry in the northern hemisphere, i.e., in Europe itself and in the newer extensions of European, or Western, civilization in North America and the Soviet Union. It has acquired immense importance, first in Europe and more recently in the world as a whole, both as a source of new ideas which challenge traditional beliefs and as a stimulus to the development of new technologies which disrupt traditional institutional arrangements.

Modern science is distinctively European in origin, despite the varied contributions of non-European civilizations in the distant past. It was in Europe, and only there, that the various component parts of science were put together in a way which gave science its unique modern effectiveness. Modern science is also distinctively European, or more broadly, Western, in its contemporary cultural status, despite the contributions of scientists from historically non-Western lands, for such scientists, along with many of their countrymen, tend to be highly Westernized in diverse ways which go far beyond the scope of their scientific activities, and are thus not genuine representatives of non-Western cultures. At the same time, science is justifiably universal in its claims; processes of nature are presumably the same in all countries regardless of local variations in cultural values. The dual status of modern science as Western in its cultural background and universal in its relevance has made it a powerful agency in the recent and continuing rise of Western culture to a position of world dominance.

Modern science has become internally differentiated, in complex and changing patterns, into numerous physical, biological and social "disciplines," or "sciences" or "branches" of science as they are variously called, and, at a more microscopic level, into sub-disciplinary "fields" or "areas" of specialization. Some disciplines, e.g., physics and astronomy, are as old as science itself, have produced impressive systems of knowledge, have served as models for less highly developed sciences, and thus most clearly exemplify the scientific process. At the other extreme are disciplines such as sociology which, although reasonably recognizable as "sciences," are comparatively young, have only relatively modest achievements to their credit and have tended to imitate other sciences rather than to serve as models for imitation by others. Such disciplines are marginal as exemplifications of science, even though not intrinsically less scientific or less important than those which provide richer sources of material for illustrating the scientific process.

Sociology, mentioned above as a branch of science, is also a discipline which includes the study of science among its component specializations. The sociology of science, as this specialization is called, is related to various other aspects of the study of science, including the philosophy and the history of science, and to various other fields of sociology, especially the sociological study of institutions and professions.

The sociology of science is a potentially important area of sociology, not only because science is important in the contemporary world, and because sociology is a branch of science, but also because science is unique among social phenomena in ways which make it quite difficult to classify in terms of sociological concepts, and which thus provide a challenge to the conceptual framework of sociology. The sociological uniqueness of science centers around one feature in particular, which pertains to the relationship between science and the natural environment.

It is commonly recognized that the natural environment does not directly determine the behavior of men, but rather merely imposes certain limitations on the range of possible behaviors, permitting variation within these limits. For example, the climate of the Arctic requires that people living there make some effective arrangements for protection against winter cold, but there are quite different arrangements which may serve this function, and the climate does not by itself determine which one a society will adopt.

It is also commonly recognized that modern technology has released men, to some extent, from limitations which the natural environment formerly imposed. For example, it is now much easier than it once was for people to obtain food produced at distant locations, and men in affluent societies are thus not limited in their choice of food, to the extent that they once were, by local food-producing capacities.

These familiar aspects of the relationship between men and their natural environment provide a background against which the central unique feature of science as a social phenomenon stands out clearly. Scientists characteristically seek to surrender their freedom of choice with respect to their interpretations of the natural world, and to have these interpretations determined precisely and completely by the natural environment as it is observed under certain special ("controlled") conditions.

This is a utopian ideal which is never perfectly realized. There is always more than one way in which a set of facts may be interpreted, and therefore always some potential freedom for scientists to choose between alternative interpretations. However, scientists characteristically do not want such freedom, but rather regard it as a problem, and seek to perform experiments which will refer imperfectly answered questions back again to nature for further clarification. Scientists *do* want freedom from control by society, but only in order that they may submit more fully to "control" by nature.

It is true that science is closely related to technology, in which the emphasis is on man's control of nature, but this does not affect the scientific emphasis on nature's control of man. It is also true that the control of man by nature which is sought in scientific activity is drastically limited in two ways: it involves control only over man's beliefs about nature itself, and it involves control by nature as observed only under certain special conditions. However, these qualifications should not be permitted to obscure the fact that, within the specified limits, science involves an emphasis on the ideal of its own complete subordination to nature. Science has, furthermore, moved toward this extreme condition, to an extent which makes it unique among social phenomena in this respect. Scientists in their research are, in effect, asking questions of nature, and they commit themselves in advance to accept whatever answers nature may give, no matter what these answers may be.

The special dependence of science on the natural environment means that the content of scientific knowledge is determined, in principle, by forces or conditions which are beyond human control. Society may encourage science, or discourage it, or make it impossible. If society does encourage science, it may stimulate developments in certain disciplines more than in others. However, society cannot, in principle, determine the contents of scientific knowledge, because these are to be determined by observations of nature. To the extent that social values intervene to require certain formulations and to prohibit others, regardless of scientifically-relevant observations of nature, the resulting formulations involve something other than "science."

While the uniqueness of science makes it an especially interesting topic of sociological inquiry, the narrow range of situations in which science has flourished and the instability of its relation to society make it difficult to study sociologically in certain important respects. The status of modern science as an outgrowth and an aspect of one civilization rules out the possibility that the emergence of modern science in the West can ever be compared with any other independent emergence of large-scale science. It also rules out the possibility that modern science can be studied in diverse independent manifestations in radically different cultures, in the way in which universal but culturally variable institutions such as the family can be studied. Furthermore, the recency of the appearance of large-scale science makes it impossible to acquire the kind of perspective which a long historical record would permit, and precludes, for the present, any definitive studies of the long-range effect of large-scale science on society. These various circumstances together impose drastic limitations on the opportunity which sociologists may have to study science at a macroscopic level on a systematic comparative basis.

In one sense, the narrow cultural distribution of science is merely a consequence of the way in which science has been defined. *Knowledge*, of one sort or another, is found in every culture. Science is merely one particular kind of knowledge which happens to be prominent in our own particular culture. Thus, if we compare the cultural distribution of science with that of the family, and say that the family is "universal" (or nearly so) while science is uniquely associated with one particular civilization, our comparison ignores a distinction between two levels of generality: we are comparing a universal institution, the family, *not* with "knowledge" as another universal institution, but rather with one particular localized variant of the latter. However, this approach may be justified on the ground that science is drastically different from all other forms of knowledge, in ways which are of great sociological importance, and which have served as a basis for recognizing the sociology of science as an area of specialization distinct from the sociology of knowledge in general.

In fact, a comparison between science and the family is particularly revealing because these phenomena are related to society in sharply contrasting ways. The family has been an important and stable component of traditional cultures, while assuming such different forms in different traditional cultures that there is an immensely rich body of data for cross-cultural comparisons. In our own modern society, the family persists, and has acquired some unique features. It may acquire radically new features in the future, under the impact of new developments in biology. However, its unique features have been consequences, rather than causes, of the more general uniqueness of Western culture itself, and under modern conditions the family has generally declined in importance. Science, by contrast, is a comparatively new phenomenon, uniquely emerging in the West, spreading around the world not through "diffusion" into other cultures in the usual sense, but rather as part of the destruction of other cultures in their traditional forms, and reacting disruptively upon traditional institutions in the West as well. It is paradoxical that these features of science which give it a special importance in the modern world also make it more difficult to study within a sociological frame of reference than are traditional institutions such as the family.

Much of the research which has been done in the sociology of science involves comparisons of scientific activities under different conditions: e.g., in different nations, different historical periods, different organizational contexts. The fact that large-scale science is distinctively Western and recent in origin inevitably limits the scope of such comparisons. It is therefore appropriate that we should seek to broaden our perspective by making comparisons not only between different manifestations of science, but also between science and other phenomena which are similar to it in important ways. This will be done in the following chapters, in the context of an attempt to classify science in relation to sociological concepts.

Chapter 2

THE CONCEPT OF SCIENCE IN SOCIOLOGICAL PERSPECTIVE

The Scientist, the Sociologist, and the Sociology of Science

Sociologists generally assume that they cannot expect to obtain an adequate understanding of a social phenomenon merely by accepting the observations and intuitive interpretations of sociologically unskilled participants. Such participants may make an important contribution as "informants," i.e., as sources of information, but not as experts competent to assess the validity of sociological information or its theoretical relevance. Thus, sociologists do not accept primitive people as experts on primitive society, or ordinary Americans as experts on American society, even though such participants may provide information about their respective societies which sociological experts will find useful.

There are at least two reasons why personal familiarity with, and participation in, a social phenomenon does not ordinarily suffice to qualify a person as an "expert" on that phenomenon from a sociological standpoint:

1) Participants are likely to be misled, unless they take appropriate measures to avoid this, by experiences which expose them unequally to different aspects of the phenomenon in which they are participating. Such differential exposure may result from the individual participant's personal status, and from the unequal visibility of different aspects of the given phenomenon itself. Thus, a citizen analyzing his own society is likely to be biased by experiences reflecting his own age, sex, ethnic background,

socioeconomic status, and geographic location, and by his inability to witness directly the complex institutional arrangements of the society.

2) Participants are likely to have a biased exposure not only to various aspects of the phenomenon in which they are participating, but also to other phenomena which provide potential bases of comparison. They also often lack a familiarity with technical concepts which facilitate comparative analysis. Thus, most people have neither a personal exposure to a wide variety of societies nor familiarity with sociological concepts pertaining to the classification of societies, and are thus poorly prepared to analyze systematically the similarities and differences between their own society and others.

Scientists, as participants in science, are likely to be biased in their descriptions of science in the same general ways in which other people are biased in their descriptions of phenomena in which they participate. A scientist analyzing science is likely to be biased by experiences which are peculiarly characteristic of his own era, his own country, his own field of specialization, and his own context of employment, and by his inability to witness directly in any systematic way the long-range trends and the complex networks of relationships which science involves. He is also likely to lack the ability to analyze in a competent way the similarities and differences between science and various other social phenomena with which it may be compared: e.g., between science and religion as forms of knowledge, between science and art as forms of cultural creativity, between science and medicine as occupations or professions, between science and the family as social institutions.

To some extent, different participants tend to acquire different biases depending on the contexts in which they participate. To the extent that this is true, the problem of participant-bias can be overcome by taking into account the differing perspectives of participants who are differently situated. For example, if a scientist's conception of science tends to be influenced by the particular problems and methods of his own field of specialization, bias arising from this source can be corrected by consulting scientists in a broad range of disciplines. However, such a procedure will be ineffective in dealing with biases which are more widely shared, which reflect the general perspective of the participant as such, as distinct from the special perspectives of those who participate in certain particular ways. To compensate for these more general biases, we must take into account perspectives which go beyond that of the participant; in particular, the perspective of the sociologist who can interpret what he observes in terms of concepts developed through comparative studies of diverse social processes and structures.

There are many possible ways in which the preceding remarks may be misunderstood, and some explicit disclaimers may therefore be appropriate. I am not suggesting that scientists generally know less about science than most other people know, or that participation in science reduces a person's capacity to understand it. I am not denying that scientists may serve usefully as sociological informants about science even when they are not sociological experts on it, or that they may overcome the limitations of the participant's perspective on science in various ways, as, for example, by acquiring an understanding of sociological concepts relevant to the analysis of science. I am not ruling out potential contributions to the understanding of science from disciplines other than sociology, or claiming that a sociologist who has little understanding of science is capable of analyzing it competently. I would agree that ideas should be evaluated in terms of their intrinsic merits, regardless of their source; that ideas should not be blindly accepted because they come from recognized "experts" or blindly rejected because they do not. These qualifications do not, however, negate the points which have been made here: that familiarity with science through participation in it does not, by itself, provide sufficient basis for a competent analysis of science as a social phenomenon, and that such an analysis belongs within the disciplinary domain of the sociologist.

Unfortunately, the condition of his discipline leaves the sociologist less adequately prepared for this than he is for various other tasks. Difficulties arising from the narrow cultural distribution of science have been mentioned in the preceding chapter. In addition, the sociological analysis of science is complicated by two aspects of the relationship between sociology and science: (1) by the status of sociology as itself a branch of science, and (2) by certain peculiarities of science as a topic of sociological inquiry.

Sociology is a branch of science, even if a rather peripheral branch in terms of its present state of development. The sociological investigator is thus inherently a participant in science. This is not necessarily a disadvantage in all respects from the standpoint of the sociological study of science; the potential value of "participant observation" is well known. However, we do not have here an ordinary sort of participant observation, in which the observer becomes a participant temporarily to facilitate a particular research project, or utilizes for research purposes some accidents of his personal background which have made him a participant. The sociologist who studies science as a social phenomenon is not merely an observer who happens, by design or by chance, to be a participant in science as well. Rather, his sociological observations of science are themselves intrinsically acts of scientific participation. It is thus harder for the sociologist to make the crucial distinction between the observer's and the participant's standpoints with respect to science, than for him to make this same distinction with respect to many other social phenomena.

The sociologist may achieve a reasonable degree of objectivity with respect to science, despite his intrinsic involvement in it, by employing, in the analysis of science, various sociological concepts which have been applied to other phenomena. However, science is quite different from most other objects of sociological inquiry, and familiar sociological concepts are thus only imperfectly applicable to it. This state of affairs, like the dual status of the sociologist as both participant in science and observer of it, provides certain potential advantages and opportunities. It means that science may constitute a valuable "test case" for the application of familiar sociological concepts and generalizations. It also means, however, that the task of the sociologist in applying the concepts of his discipline to the analysis of science is more difficult than it would be if science were merely what one might call an "ordinary" social institution.

Each of these two problems in the relation between sociology and science has tended in a different way to narrow the scope of the sociological study of science. On the one hand, the status of sociology as a branch of science, and of sociologists as scientists, encourages sociologists to accept uncritically, for *theoretical* purposes, the conception of science which scientists generally employ for *practical* purposes in connection with their work. This in effect removes the question of how science is to be defined, from the range of questions to be answered within a sociological frame of reference. On the other hand, the conceptual framework of their discipline has encouraged sociologists to overlook or de-emphasize those aspects of the social organization of science which cannot be effectively interpreted in terms of such sociological concepts as "social system," "institution," "occupation," and "profession."

These limitations on the scope of sociological inquiry into science are understandable in view of the problems with which that branch of sociology is confronted, but are nevertheless arbitrary and unwarranted in terms of the highest standards of sociology as a scientific discipline. In terms of these standards, it is not enough that the sociology of science should include inquiries into "the relation between science and society" and into "the social organization of the scientific community," with the concept of science itself defined in advance by scientists other than sociologists, or by philosophers of science. Instead, the sociology of science should include as one of its tasks the formulation of a sociologically adequate conception of science itself, reflecting the viewpoint of the sociologist as observer of science, rather than the viewpoint of the scientist (including the sociologist) as participant. On the other hand, it is important to avoid the arbitrary assumption that science can be readily subsumed under familiar sociological categories. Instead, sociological concepts should be applied to science only hypothetically and tentatively, with full allowance for the possibility that such concepts may sometimes obscure rather than illuminate the distinctive features which alone make science worthy of our attention.

We may now review some familiar conceptions of science within the framework of these guidelines.

The Conception of Science as a Method

In his work the scientist pursues goals through use of methods. The conception of science which he is likely to acquire from his own participation in it involves goals and methods which are generalizations of his own. The recognized goal and methods of science have changed over time; scientists of today, for example, differ from some of their distinguished predecessors in that they generally state their scientific goals without reference to God. There are also differences of opinion today concerning the precise way in which the goals and methods of science should be stated, and considerable attention has been devoted to this question by philosophers of science and by others. However, the various modern formulations of scientific goals and methods agree substantially on certain basic points. The goal of science, as commonly recognized today, involves the acquisition of systematic, generalized knowledge concerning the natural world; knowledge which helps man to understand nature, to predict natural events and to control natural forces. The scientific method employed in pursuit of this goal involves use of previously accumulated knowledge to construct general theories or systems from which testable hypotheses can be derived, and the testing of such hypotheses via quantified observations under controlled conditions.

It is true that individual scientists, and various scientific research teams and institutes, do have scientifically relevant goals which they pursue through specific methods. It is understandable that scientists, analyzing science as a whole on the basis of their own experiences and activities, should tend to define it in terms of goals and methods which are generalizations of their own. It is understandable, also, that sociologists, as participants in science, should commonly accept such a definition. On the other hand, it is also appropriate that sociologists of science should examine such a conception critically, in terms of its adequacy as part of the conceptual framework of their specialization. The sociology of science would have no justification for existence if it had nothing to tell us about science, beyond what scientists themselves already know. The potential contributions of the sociology of science can be maximized by defining science in a way which reflects not the perspective of the scientist (including the sociologist) as a participant in scientific activity, but rather the perspective of the sociologist as an observer of science.

From this latter perspective, the conception of science as a "method employed in pursuit of goals" is subject to criticism on the following grounds:

1) A "goal" as usually understood is an outcome toward which people strive, or more generally toward which the internal functioning of a system is directed. The term does not ordinarily encompass outcomes toward which movement occurs in accidental ways, e.g., through mechanisms comparable to the "random variation" and "natural selection" of biological evolution. Goals in this sense are meaningfully attributable to individual people, including scientists, and to certain groups and organizations, including scientific research teams, and scientific institutes. However, science as a whole is a process which transcends particular scientists, research teams, and institutes. The meaning of a state-

ment attributing a "goal" to such a process would require clarification. Science produces certain outcomes, and some of these outcomes are goals of individual scientists and research teams; but it does not necessarily follow that any outcomes of science must be classified as "goals" of the scientific process as a whole.

2) Even if a goal is attributed to science, it does not necessarily follow that science must be *defined* in terms of movement toward that goal. It is entirely possible that the most significant aspects of science involve movement, over a long time span, in directions which have not been intended or recognized by scientists generally, and which have emerged accidentally, even if there has also been movement in directions which may be identified as corresponding to a "goal" of science.

3) Finally, even if science is to be defined as a process of moving toward a goal, it does not follow that science thereby becomes equivalent to a "method." A method is a process employed deliberately in pursuit of a goal. Methods are used within particular scientific inquiries. However, the concept of method cannot reasonably be applied to some important types of events through which the findings of different inquiries are interpreted and integrated by the scientific community as a whole. Sometimes, for example, two different approaches to a scientific problem compete with each other for a long time, each supported by a different group of scientists, and no known experiment can decisively resolve the issue because each of the two approaches suggests a different type of experiment and a different way of interpreting the results. The complex adjustments through which public opinion within the scientific community comes to crystallize around one such approach while abandoning the other characteristically cannot be reduced to a "method," but are nevertheless an important component of the scientific process as a whole.[1]

The preceding considerations suggest that the concepts of goal and method be recognized as appropriately applicable to science at relatively microscopic levels only. When we observe what is done by a particular scientist or research team, working on a particular problem, we see methods used in pursuit of goals. When we seek instead to analyze science macroscopically, taking into account not merely what happens within particular research projects but also the integration of findings of many such projects in different disciplines over centuries, the concepts of goal and method appear to lose their relevance.

The Conception of Science as a Social Institution

Sometimes the term "institution" is used as synonymous with "organization," as when schools, prisons, churches, and business corporations are labelled as "institutions." At other times the term is used as synonymous with "social norm," "custom," or "tradition." When the term is used by sociologists not merely as a synonym for such other, well-established terms, but rather with a distinctive and theoretically relevant meaning of its own, it is commonly used to refer to a social pattern or arrangement which is integrated with other such patterns, performing functions for society and receiving societal support in return. I shall employ the term here only in this latter way. Thus, religion, education, family, economy and polity may be recognized as "institutions," each performing important and familiar functions for society.

Application of this concept to science is impeded by the absence of clear and consistent functions which science can be recognized as performing for society in exchange for the support it receives. In fact, the particular course of development which science will follow is largely uncontrollable and unpredictable. It is true that scientists "control" the situations in which they make their observations, and that they "predict" what they will observe in these situations, but this does not involve either controlling or predicting the direction in which science itself develops. The course of scientific development is "uncontrollable" in that the scientist commits himself in advance to accept whatever answers nature may give to his questions and those of his colleagues, and although men must interpret nature's answers, there are major limitations to the freedom which they have in making such interpretations if the results are to be recognized as scientific. The course of scientific development is unpredictable because we cannot tell, at any given time, how nature will answer the questions which scientists will ask in the future, or even what these future questions will be. In these respects, science resembles an adventure more than an institution. Furthermore, the adventure is one which involves the entire society of which science is a part. Science has an unstabilizing effect on other institutions, and thus cannot be viewed as part of a stable institutional pattern. A society with major emphasis on science is invariably a society undergoing drastic change which presumably affects all its institutions.

This dynamic aspect of science means that "functions" which may be claimed for science are likely to be controversial with respect to their functional status, so that what science does *for* society is not clearly distinguishable from what science does *to* society. It also means that the functional significance of science for society may vary considerably according to time and place, so that societal functions of science, even when they can be clearly identified, are likely to have only temporary and local significance. The problems inherent in any attempt to identify universal functions for science are nicely illustrated by the contrast between Bronislaw Malinowski's classic statement that the functional distinction between religion and science is clear even in primitive societies,[2] and the recent description of the scientific elite in our own society as "the new priesthood."[3]

The difference between science and typical institutions in these respects is profound. In every society we find economic, political, religious and educational institutions, sometimes highly developed, and sometimes relatively undeveloped. These institutions have characteristic functions which are generally recognized, and in some cases important latent functions also. Although such institutions may undergo important changes under some conditions, they may remain essentially unchanged for many years under other conditions, each performing its distinctive functions on a stable, continuing basis. Science, by contrast, has not emerged on a large scale except in our own Western civilization in recent centuries, and in that civilization it has been primarily a disruptive rather than an integrative force, transforming society drastically in ways which may ultimately prove to be either enormously beneficial or disastrous, but which in any case do not involve a stable relationship between science and society.

Perhaps the closest reasonable approach that can be made to an analysis of science in "institutional" terms in the present sense would involve a conception of science as a specialized local outgrowth of traditional "cognitive" institutions. In this respect an analogy between science and a similar outgrowth of traditional economic institutions may be appropriate. In traditional societies, and in some modern societies, prices have been largely controlled by custom or by official regulation, just as knowledge has been similarly controlled. However, in certain times and places in the Western world in recent centuries the market has largely escaped from such control, with prices coming to be determined largely through the autonomous, impersonal mechanism of "the law of supply and demand," just as knowledge has similarly escaped from authoritarian control and acquired, in science, its own autonomous mechanisms of development. The impact of the free market on society has been so drastic that major efforts have been made to reimpose heavy controls on its operation, just as the impact of science on society has stimulated renewed efforts directed toward social control of science. In these cases, we are dealing essentially not with "institutions" in the present sense, but with specialized outgrowths of institutions which have largely broken away from social controls which previously prevailed, upsetting the systematic arrangements which formerly related them to other aspects of society.[4]

It is reasonable to classify science as an "institution" if one is referring only to science in particular modern societies over relatively brief periods of time, or if one is defining the concept of "institution" in a relatively broad way which does not imply a stable relationship with the larger society. However, given a definition of "institution" which does imply such stability, that concept is only imperfectly applicable to science as a whole.

The Conception of Science as an Occupation

Even in societies in which great emphasis is placed on science, those people who participate directly in the scientific process never constitute more than a small percentage of the total adult population. The overwhelming majority of people everywhere are "non-scientists" who do not have much understanding of science or much interest in participating in it.

For those relatively few people who do participate directly in science, such participation generally constitutes a major personal commitment. It is something that they want to do, a source of enjoyment, and to the extent that they are "successful," a source of esteem. It is characteristically *not* something that they do *merely* as a means of earning a living: in fact, men engaged in pure scientific research have often been able to earn more money if they abandoned that activity, and used their skills instead in "applied" research directed toward the solution of practical problems.

Science, then, involves activities by a small number of strongly committed people, rather than a large number of people whose interest in science is casual. In two respects this state of affairs has facilitated scientific progress:

1) A significant contribution to science normally requires an investment of time and energy, both in educational preparation and in research, which a merely casual interest in science would not suffice to bring forth. The strong personal commitments which scientists commonly have to their scientific work enables many of them to make scientific contributions which would not otherwise be possible, and which, specifically, could not be made even by a very much larger number of investigators whose involvement in science is relatively casual.

2) Science would hardly have been able to progress very far

if new scientific ideas had to be understood and accepted by the population of a society in general, including many people with only a casual interest in science, before these ideas could be regarded as scientifically established and utilized as a basis for further scientific work. The fact that scientific activity is localized in a specialized segment of the total society makes such widespread diffusion unnecessary. Instead, new ideas need only be accepted by a relatively small number of scientific workers, and with acceptance at this level, a new idea may be incorporated into the prevailing body of scientific knowledge regardless of what the "general public" may think of the matter, if indeed the general public has any awareness of it at all. Thus, the localization of scientific activity within a small part of society helps to make it possible for science to be somewhat autonomous in relation to societal culture, to develop in directions which may not be completely consistent with prevailing cultural emphases in the society as a whole, and, to progress more rapidly than it could if it depended upon wider understanding and acceptance among relatively uninterested people.

The status of science as a specialized activity of a few people suggests that science might be usefully conceived as part of the "division of labor in society," as an occupational role. The strong personal commitments which scientists have to their work, together with the high educational qualifications which they generally have, suggests that science might be classified more specifically as an occupational role of the "professional" type. Thus, we might regard scientists as essentially similar to factory workers, secretaries and janitors in that they earn a living by performing specialized services, and similar to physicians, lawyers and university teachers not only in this latter respect but also in terms of the features which distinguish "professions" from ordinary non-professional "jobs."

However, although the role of "scientist" does resemble, in important ways, various occupational roles in general and various professional roles in particular, there are nevertheless some important ways in which the role of scientist simply fails to fit the usual conception of an "occupation" or "profession." Furthermore, the divergences between the role of scientist and ordinary occupational and professional roles are not merely accidental or random: rather, these divergences are systematically related to each other, and reflect important features of the scientific process.

Major differences between the role of "scientist" and ordinary occupational roles include the following:

1) To a large extent, scientific research has been carried on by people for whom this is not a full-time gainful occupation. Contributors to science who have not been gainfully employed as scientists on a full-time basis may be divided roughly into two categories:

a) First, there have been some very important contributions to science, and numerous minor contributions, by investigators who have not been employed as scientists at all, or whose scientific employment has been merely incidental. Among those who have made important contributions to science while not being gainfully employed as scientists we may mention Anton van Leeuwenhock (1632-1723), janitor of the town hall of Delft and discoverer of microbes; Benjamin Franklin (1706-1790), who achieved fame as a scientist and also as an inventor, author, and statesman; Gregor Mendel (1822-1884), Augustinian monk whose discoveries came to serve as the foundation for the scientific study of heredity; and Albert Einstein (1879-1955), who was an examiner in a patent office when he formulated the special theory of relativity. And, in recent times, certain kinds of scientific research which can be undertaken by individuals on a part-time basis without expensive equipment or elaborate formal training have attracted relatively large numbers of amateurs. For example, much of what is known about patterns of bird migration and distribution represents the accumulated result of observations by numerous amateur ornithologists.

b) Secondly, there are scientists who are gainfully employed in occupations related to their scientific skills, but who nevertheless do not engage in pure scientific research on a full-time basis. This category would include those who make contributions to science, but who spend part of their working time in such related-but-different activities as teaching, administration, applied research (as distinct from pure scientific research), textbook writing, consulting, and professional practice in medicine.

These situations in which scientific activity has been undertaken on some basis other than that of a full-time gainful occupation, are not merely minor exceptions. The list of men who have made scientific contributions while not being gainfully employed in any capacity related to their scientific work includes some of the greatest scientists of all time, and more specifically some of the greatest scientists in each of the centuries since modern science began. And, a very large part of contemporary scientific research is undertaken by men who do this on less than a full-time basis, who combine pure scientific research with one or more of the other related activities enumerated above. We thus have one important way in which the role of scientist diverges from the usual conception of an occupational role.

2) Scientific roles differ from ordinary occupational roles not only in the fact that they commonly involve much less than a full-time gainful occupation, as discussed above, but also in another respect. Most occupational roles, including professional roles, entail some kind of service which is performed, for which the role incumbent receives compensation. In other words, there is some pattern of exchange. Soldiers are paid for fighting; janitors for cleaning buildings; physicians, lawyers and teachers for providing appropriate service to their patients, clients and students; grocers for providing food to customers. A pattern of exchange is present even in cases of "exploitation" of labor, such as slavery, although the exchanges in such cases are weighted to favor one side at the expense of the other.

Scientists who are also teachers, administrators, technologists and consultants may perform clearly identifiable services in these latter capacities, for which they are paid. But what about the scientist insofar as he engages in the pure scientific research which constitutes the central core of the role of scientist as ideally conceived? There are special problems involved in identifying the "services" which pure scientists perform, comparable to the services provided by incumbents of more ordinary occupational roles.

The outcome of pure scientific research is likely to be drastically unpredictable: not merely in the sense in which the achievements of a soldier, physician, lawyer or teacher may have an unpredictable ranking on a undimensional scale between the extreme poles of "complete success" and "complete failure," but in the more fundamental sense that there may be an unlimited number of possible varieties of scientific "success," along with the possibility of "failure." In addition to the uncertainty of success, there is uncertainty as to the kind of success which will be achieved, if any. Further, each of these uncertainties may appear at each of two levels: there may be uncertainty concerning both the type of theoretical structure which will emerge as a solution to a problematic situation in science, and the practical, technological implications of this structure. But these various uncertainties are not the only problem we have, when we seek to identify the "services" which scientists may be conceived as providing, in exchange for whatever support their purely scientific research receives. Practical benefits from pure scientific research are not only unpredictable; they are also often long delayed, and when they do materialize they do not necessarily favor those who supported the research as distinct from those who did not.

Hence, the usual model according to which the incumbent of an occupational role performs "services" for which he is compensated, is quite difficult to apply to the role of scientist in its purest form. Scientists do commonly perform clearly identifiable services for which they are paid, but these services most often involve teaching, applied research, administration, consulting and other activities which are peripheral rather than central to the "scientist" role. Scientists in connection with their pure scientific research do perform services for each other, by exchanging ideas and information, but these exchanges are strictly within the scientific community and do not provide scientists with whatever resources they may need from outside their own ranks.

We thus have two features of the role of scientist which render the usual conception of "occupational role" only imperfectly applicable to it. First, the role of scientist is only imperfectly differentiated from other occupational roles, in the sense that scientists are usually gainfully employed in other occupations in addition to (and in some cases totally in place of) being gainfully employed in pure scientific research. Secondly, the role of the scientist as an occupational role is imperfectly integrated with other roles, in the sense that the services provided by the scientist, in exchange for whatever compensation he receives for his purely scientific work, are commonly unclear. These two features are both part of a single pattern: scientists commonly have gainful employment outside of pure scientific research, and this gives them a clearly defined place in the occupational system, which the scientific role in its purest form fails to provide. This pattern has an important function for science. The fact that much of the scientific process takes place outside the normal framework of the occupational system of society facilitates the independence of science from the thoughtways of the societal culture, and from the limitations entailed in the expectation that achievements for which one is paid should have a recognizable practical utility.

The Conception of Science as a Profession

Scientific activity does possess to some extent the characteristics of an occupational role, even though, as I have pointed out in the preceding section, this activity is imperfectly differentiated from, and imperfectly integrated with, other occupations. To the extent that science is an occupation, we can meaningfully consider the possibility that it may be classifiable as an occupation of the "professional" type. In investigating this possibility we shall find still another illustration of the failure of science to fit neatly into familiar sociological categories.

There are differences of opinion among sociologists concerning the way in which the concept of "profession" can best be defined. Such differences are reflected in the fact that some sociologists describe science as a profession,[5] while others distinguish between a profession and a science.[6] For present purposes it will be desirable to employ a definition which is relatively broad, which may be generally acceptable to sociologists as a partial definition even though some will wish to add further specifications. In attempting such a definition I shall distinguish between features of professional work itself, and features of the organization of professional activity.

A profession may be recognized as having two essential features which pertain to the work which is involved: 1) professional people perform services which require highly specialized knowledge and skill, and 2) when professional services are performed, the outcome depends not only on the competence with which the professional work is done, but also to a considerable extent on situational factors beyond the control of those who provide the services. Physicians, lawyers, and university professors are characteristically professionals in this sense. All three of these occupational roles entail high levels of knowledge and skill. In all three cases, also, the outcome of the relevant occupational activity depends heavily on situational factors: there are many patients whom no physician can save, many clients whose interests no lawyer can defend with complete success, and many students whose academic progress is impeded by factors beyond the control of any professor. Elementary school teachers are usually only semi-professionals in the present sense, because their skills are not very highly specialized. Commercial airplane pilots are similarly semi-professional, but in a different way: they utilize highly specialized knowledge and skill, but in their work the part played by uncontrolled situational variables has been reduced to such an extent that the safe and successful conclusion of each flight is taken for granted as a normal result of the application of proper knowledge and skill.

This conception of a "profession" has two important implications concerning the social organization and control of professional activity:

1) Competent professional activity requires independent judgment by the individual practitioner, based on his general professional knowledge and his understanding of the unique conditions prevailing in the immediate situation in which this knowledge is to be applied. Hence, if the professional man is to operate effectively, the system of social control within which he operates must allow for considerable autonomy for him in his work, and he must be properly motivated to utilize this autonomy in ways which are appropriate to his professional responsibilities.

2) Competent evaluation of professional work requires that the effects of this work be distinguished from situational effects beyond the professional practitioner's control. The importance of such a distinction is illustrated by the possibility that a physician who saves the life of only one patient out of one hundred may thereby be performing excellently *if* he is treating a disease which is ordinarily survived by only one out of one thousand of its victims. Only professional colleagues can be counted on to make proper allowance for situational factors, and hence only they can be considered generally competent to evaluate professional activity.

Competent professional organization thus requires considerable autonomy both for the individual professional man in the performance of his work, and for the profession as a whole in the evaluation of performances by its members.

Scientific work, insofar as it constitutes an occupation, is clearly an occupation of the professional type as here defined. In fact, science appears to satisfy the stated criteria of a "profession" more fully than these criteria are satisfied by most occupations which are recognized as professional. On the other hand, science differs importantly from most professions, in certain organizational features:

First, there are no licensing procedures or formal educational requirements for participation in science, as there are for work as a physician, lawyer or teacher. There may be, and usually are, such requirements for *employment* in scientific capacities, e.g., for employment as a university professor in a scientific discipline, for which a Ph.D. degree may be essential. However, as I have pointed out above, employment as a scientist is not essential in principle for being recognized as a scientist, or for participating in the process of science, even though it may be essential from a practical standpoint in many particular cases. The journal which published Einstein's classic paper introducing the special theory of relativity was acting in accordance with strong scientific tradition when it accepted this paper on its merits, and thus implicitly accepted Einstein himself as a legitimate participant in the scientific process, regardless of the fact that his employment at the time was only as an examiner in a patent office.

Entrance into a profession such as medicine, law, or university teaching is commonly subject to official control by governmental or other non-professional authorities, although this control is actually exercised to a large extent by professionals who act on behalf of the authorities, who determine which candidates meet the requirements which authorities have set — and who may participate importantly in setting the requirements themselves. Thus, an MD degree may be legally established as prerequisite for entrance into medical practice, although the organized medical profession itself may in effect decide to whom such degrees are to be awarded, and may be largely responsible for maintaining the MD as a legal requirement. The absence of formal licensing requirements and formal educational requirements for entrance into the scientific profession, makes the situation in that profession different from the situation in others, in two ways. On the one hand, it means that entrance into the scientific community is in principle determined *exclusively* by the scientific community itself, without any involvement in this determination by outside authorities. In this respect, science actually comes closer to the ideal model of a "profession" than do most other professions. On the other hand, the absence of formal admission requirements also means that ordinarily the scientific community cannot (and would not seek to) call upon the government for assistance in curtailing the activities of pseudo-scientific "quacks," who are not members of the scientific community, in the same way in which the medical and legal professions can often protect themselves from competition by unqualified outsiders. To imagine a comparable situation in medicine or law, one would have to suppose that anyone who wished to do so could practice medicine or law as he pleases (just as anyone is free to attempt scientific research), and that those who practice successfully would then be recognized as members of the appropriate professional group.

Secondly, the scientist as pure researcher has no "clients" comparable to the lawyer's clients, or the physician's patients, or the teacher's students or their parents. Pure scientific researchers may have employers and sponsors, but the unpredictability of scientific discoveries, the unpredictability of the practical applications of such discoveries, and the long intervals of time which often elapse between discovery and application, mean that employers and sponsors often cannot be classified as clients for whom pure scientists perform clearly identifiable services in the usual professional sense. Furthermore, the scientific community has no "code of ethics" pertaining to the responsibilities of scientists to non-scientific sponsors, comparable to the codes of ethics of the medical and legal professions which protect physician's patients and lawyer's clients from "malpractice." In fact, the closest approximation to a "client" of the pure scientist, in these respects, would be the scientific community itself. That is, pure scientists, regardless of their employment or sponsorship, seek primarily to provide services for other scientists in their own and related areas of specialization, and are governed by a code of ethics or normative system which is primarily concerned with the regulation of these services within the profession in distinction from services by the profession to outsiders.

The two forms of divergence from the usual "professional" role which have been noted here are closely inter-related. The ordinary professional man — physician, lawyer, professor — provides services to relatively helpless and "ignorant" people who cannot easily protect themselves against "malpractice." Furthermore, the contexts in which these services are performed make it relatively difficult for other professional colleagues, or other non-professional authorities, to provide supervision effective enough to prevent such "malpractice." Hence, the importance of controlling admission to the professions in question, in terms of educational and ethical qualifications, even though, in practice, the admission requirements may serve other functions, such as limiting competition for established members of the given profession. By contrast, the "clients" for whom a scientist directly performs services are essentially other members of the scientific "profession" who do not need the kind of protection against malpractice which clients of ordinary professionals need. Each "service" which a scientist performs for his colleague-clients, in the form of a research report submitted for publication or distributed in some other manner, can be relatively easily evaluated by these recipients of his work in a way in which a client of a physician could not normally evaluate the latter's services.

These differences between science and "ordinary" professions which have been noted here can largely be understood as consequences of another, more fundamental feature which differentiates science from most other professions: the fact that scientific work is *creative* in the sense that each scientific contribution must be "unique."

Apart from cases of simultaneous independent discovery, and cases in which a discovery made in one cultural context is lost or remains localized and hence has to be made separately elsewhere, no two scientists can make exactly the same contribution to science. The uniqueness of each scientist's contribution is a matter of definition in the sense that a purported "contribution" which is not original and hence not unique is simply not regarded by the scientific community as a genuine contribution. This does not mean that individual scientists are "irreplaceable," or that replacement of one scientist by another changes the direction of scientific development. On the contrary, two kinds of "replaceability" are taken for granted in science. First, it is ideally expected that when a scientist reports results of his experiments, any other competent scientist, repeating the "same" experiment under the "same" conditions should obtain essentially the "same" results. Secondly, it is recognized that a discovery which has actually been made by a particular scientist would very likely have been made sooner or later by some other scientist instead, if the one who has made the discovery had not done so. However, these kinds of "replaceability" do not affect the unique quality of each contribution to science.

Some individual variation in role performance is found almost everywhere. It is usually possible to find some kind of "uniqueness" in the way in which a person goes about his job. In some cases the unique features may be trivial, as when factory workers on an assembly line have their actions closely regulated in accordance with the requirements of the production system. In other cases, and especially in professional occupations, individual variations may be much more important: different teachers have different styles, and may accordingly differ considerably in effectiveness. However, it remains generally meaningful to say that two teachers are doing essentially "the same thing" even though they may be doing it "in different ways." On the other hand, when we move from consideration of teaching to consideration of scientific research, we cross a boundary line beyond which it is no longer clearly meaningful to say that two members of the same occupational group are making "essentially" the "same" contributions in their work. The genuine contributions of a scientist consist exclusively of what may be considered analogous to "individual variations in role performance" in most other occupational roles, although of course these variations are acceptable only if they are compatible with certain features of the scientific role which remain standardized, including the requirement that appropriate scientific methods be followed.

The uniqueness of each scientific contribution marks science as belonging in the general category of culturally creative activities, along with literary, musical and other forms of artistic creativity. Science, then, is both a professional and a culturally creative activity. But typical creative activities are not professional, and typical professional activities are not significantly creative in a cultural sense, even though they may involve creativity at a personal level which leaves the culture unchanged. In this respect science represents an unusual combination of features: it differs from most creative activities, by virtue of its professional organization, and it differs from most professions by virtue of its emphasis on cultural creativity. On the one hand, the professional aspect of science means that its creative productions, in contrast to those of most culturally creative occupations, are primarily evaluated by professional colleagues of the creators, rather than by an outside clientele, and are evaluated by such colleagues on the basis of rigorous objective standards. On the other hand, the creative aspect of science appears to be largely responsible for the organizational features already noted, which differ considerably from the corresponding features of most professions. The absence of official prerequisites for entrance into the scientific role, and the frequent absence of non-professional clients for whom professional services are performed, means that in these respects science differs from typical professions and resembles, instead, such non-professional creative activities as that of the poet, who also needs no license to engage in his line of work, and who has no professional-type obligations to clients.

I have pointed out that science fails to fit perfectly the concept of a social institution, when the latter is defined as a pattern which performs functions for society and is supported by society in turn, and that science can be conceived as an "occupation" only to a limited extent. To the extent that science is an occupation, it tends to resemble occupations of the professional type, although it differs importantly from "ordinary" professions by virtue of its culturally creative character. We thus find that science is not at all easy to classify in terms of the sociological concepts which might superficially appear most applicable to it. Furthermore, the deviations of science from familiar sociological models are not minor or incidental matters: rather, they pertain to precisely those features of science which make science important and interesting.

An Alternative Conception of Science

The conception of science as a "method" used in pursuit of a "goal," which represents the perspective of the participant in science, is microscopic and subjective, and hence inadequate as a basis for a sociological analysis of science as a whole. Sociological conceptions of science as an "institution," "occupation," and "profession" are also inadequate insofar as they imply a comparatively stable relationship between science and society, with science performing certain functions or services on a relatively consistent basis. A conception of science as a cultural process avoids these difficulties. It brings science within a sociological frame of reference without necessarily assuming that science as a whole can be identified either with specific "goals" or with specific "functions." These assumptions can be avoided because the concepts of culture and cultural change do not entail either goal-direction or functionality. This point can be illustrated with examples pertaining to the linguistic aspect of culture: changes in meanings of words, pronunciation, rules of grammar, etc., often occur without being deliberately sought by anyone, and without being recognizably useful or "functional" for anyone or for any social system. By viewing science as a cultural process we allow for the possibility that science may similarly lack identifiable goals and functions. The chapter which follows will discuss the conception of science as a cultural process in some detail.

NOTES

1. It has often been assumed that conflicts between different scientific approaches could generally be resolved relatively easily by undertaking observations or performing experiments which would provide a clear basis for choosing between them. Thomas S. Kuhn, in *The Structure of Scientific Revolutions* (Chicago: University of Chicago Press, 1962), has pointed out that the process by which scientists choose between competing approaches is commonly much more complicated than this, and cannot readily be reduced to methodological rules.

2. Bronislaw Malinowski, *Magic, Science and Religion*, Glencoe, Ill.: Free Press, 1948.

3. Ralph E. Lapp, *The New Priesthood: The Scientific Elite and the Uses of Power*, New York: Harper & Row, 1965.

4. Karl Polanyi, in *The Great Transformation* (Boston: Beacon Press, 1957), discusses the development and impact of the free market economy in a way which suggests important analogies with the development and impact of science. It is presumably not entirely a coincidence that these two profoundly disruptive outgrowths of traditional institutions have both appeared prominently within the framework of the same (Western) civilization in roughly the same historical period.

5. Norman W. Storer, *The Social System of Science*, New York: Holt, Rinehart, and Winston, 1966, pp. 91-97, and Randall Collins, "Competition and Social Control in Science: An Essay in Theory-Construction," *Sociology of Education*, Spring 1968, v. 41, no. 2, 123-140.

6. Harold Wilensky, "The Professionalization of Everyone?", *American Journal of Sociology*, v. 70, no. 2, September 1964, 137-158, and Everett C. Hughes, *Men and their Work*, Glencoe, Ill.: Free Press, 1958.

Chapter 3

SCIENCE AS A CULTURAL PROCESS

This chapter presents a conception of science as a cultural process, and as a process which is also cognitive and developmental. Science as thus conceived may be distinguished from, and compared with, other phenomena which resemble it in any two of these three ways but not the third. Such phenomena include:

1) the cognitive development of the individual as he moves from infancy to maturity: a process which is similar to science in being both cognitive and developmental, but different from science in being individual rather than cultural;

2) various non-scientific cultural knowledge systems — e.g., primitive, ancient and medieval world views, and some modern philosophies — which are similar to science in being both cultural and cognitive, but which lack the intrinsic developmental character of science as a process; and

3) various processes of cultural development which are similar to science in being both cultural and developmental but which differ from science in that they pertain primarily to aspects of culture other than the cognitive.

Each of the comparisons thus indicated directs attention to a different aspect of science, and each provides a different perspective from which science may be viewed. Thus a single basic approach to the study of science will be used to generate three distinct perspectives on science, perspectives which are mutually compatible and capable of being integrated into a single more comprehensive perspective, but which differ in the insights which they provide.

Science as Cultural Counterpart of Individual Cognitive Development

Our capacity to observe the world is limited by the selectivity of our perceptions, i.e., by the selective sensitivity of our sense organs, and by the selective focusing of our attention on some aspects of the environment to the neglect of other aspects. It is also limited by our restricted locations in time and space. The cognitive development of the child involves a process of learning to correct for certain limitations on his observational capacities. Science may be recognized as an extension and counterpart of this correction process, or cognitive development process, on a cultural rather than an individual level.[1]

We may assume that the child at the start of his cognitive development is "egocentric" in the sense that he naively accepts his immediate impressions as reflections of total reality.[2] Two features of this initial egocentrism may be distinguished. On the one hand, the egocentric subject accepts his immediate observations of the world without allowing for the limitations of his perspective, without realizing that he sees the world only from one among innumerable possible perspectives. On the other hand, he fails to distinguish between those aspects of his impressions which represent genuine observations and those which reflect internally-stimulated sensations. From the standpoint of the adult, he cognitively accepts too little in one respect and too much in another: too little in the sense that he fails to take into account perspectives other than that of his own immediate situation, and too much in the sense that he fails to screen out those impressions which have an internal rather than an observational origin.

Given these two aspects of the initial egocentric condition, we may correspondingly identify two aspects of the developmental process which leads away from egocentrism:

1) The growing child implicitly learns that an object which appears small might really be large but far away; that the apparent motion of an object might be merely a reflection of his own motion; that the changing appearance of an object might be caused merely by a change in his own position relative to it; and that an object might continue to exist even when he can no longer perceive it. Later, he learns that objects appear differently to different people (rather than merely to the same person under different observational conditions), and that this applies not only to simple physical objects but to complex symbolic objects as well. (For example, a given political doctrine may appear differently from the standpoints of different interest groups.) Thus, the child learns to take into account perspectives other than that of his own immediate situation.

2) The child also learns to distinguish, even though imperfectly, between what he observes and what he imagines, dreams, or feels without observational referents. He learns, that is, to distinguish between external stimuli impinging on him from outside, and stimuli of primarily internal origin.

To a certain extent these developments occur universally among normal individuals regardless of variations in cultural context; thus, normal children everywhere learn explicitly or implicitly that the apparent size of an object is a function not only of actual size but of distance. On the other hand, some corrections for observational limitations are subject to considerable cultural and individual variation. Thus, people differ considerably in their ability to put themselves mentally in the social positions of other people, and to examine abstract problems from diverse points of view.[3] And, there are domains in which the correction process, as a process of *individual* cognition, encounters insuperable obstacles. These obstacles may impede both aspects of the correction process — both the emergence of constructs which make allowance for observational bias, and the differentiation between "observation" and "imagination."

Some corrections are technically difficult to make and to validate. A man looking up at the sky with unaided eyes and without the knowledge that has accumulated through the centuries will be likely to "see" that the moon is larger than the stars, that the sun is the brightest object in space, and that the earth is a stationary object beneath a heavenly dome. He might, or might not, be consciously aware that his view of the heavens is systematically biased by his inability to observe except from the vicinity of the earth. But even if he is aware of the possible existence of such bias, he is unable by himself to determine the precise corrections which are appropriate. Thus, Nicholas of Cusa (1401-1464) explicitly recognized that no matter where in the universe a man is located, "it will always seem to him that the position he occupies is in the motionless center, and that all other things are in motion,"[4] but this awareness of his own observational bias did not suffice to enable him to make the appropriate corrections in his interpretation of the observed universe.

When observed situations are ambiguous, people commonly fill in the gaps in their observational knowledge through imaginative speculation. The resulting speculative interpretations may then become integrated into the prevailing culture, as illustrated by various primitive, ancient and medieval cosmologies and astronomies which attribute to heavenly bodies various traits which go far beyond any which may have been observed or reasonably inferred from observations of the heavens. Such speculations, to which people may acquire strong commitments, are *not* "corrections for observational limitations;" they are a substitute for the correction process, and when firmly entrenched they constitute a major obstacle to its extension.

In a maximally pre-scientific cultural situation, the correction process is blocked by obstacles of both types cited above: by technical difficulties arising within the correction process itself, and by speculative assumptions which originate outside that process and which impose arbitrary restrictions on its scope. In such a situation, immediate observational impressions are accepted naively and mixed indiscriminately with imaginative speculations which fill in observational gaps. For example, a prescientific astronomer may accept his immediate impressions of the heavens (e.g., the impression that the moon is larger than the stars) while at the same time he fills in observational gaps with culturally-established myths which he does not carefully distinguish from his observations themselves (e.g., myths about origins of heavenly bodies or about "forces" which move them).

Science may be viewed as a correction process which has been transformed in such a way that it has successfully broken out of the bonds formerly imposed by the obstacles cited above. In this transformation, the correction process has ceased to be merely a universal activity of separate individuals, and has become, instead, a specialized cultural phenomenon. Because it is cultural, it is able to combine and accumulate relevant achievements of numerous individuals over long time spans, so that participants can draw upon a pool of discoveries and ideas contributed by their colleagues and predecessors. Because it is specialized, it is free to move ahead of the culture of the society as a whole, and sometimes in directions quite incompatible with the prevailing emphasis of societal culture. With this transformation, the correction process gains the advantages of collective activity and the advantages of specialized activity: it is burdened neither by the need for each participant to rely exclusively or primarily on his own limited resources, nor by the need for general societal consensus at each stage of development.

In thus identifying science as an extension of individual cognitive development I am not denying that science, in turn, has made possible a vast extension of individual cognitive development among many participants in scientifically advanced cultures, e.g., that modern school children can solve problems and answer questions which would have baffled the greatest thinkers of ancient and medieval times. There is a circular causal relationship here, and we should be careful to avoid interpreting science as an "extension" of individual cognitive patterns which themselves may have emerged only within a cultural framework already permeated with scientific ideas. However, such complications do not provide a basis for seriously challenging the view of science as an extension of those cognitive-development process which prevailed, either universally or locally in relevant cultures, prior to the emergence of science.

But science is not only an *extension* of individual cognitive development; it is also, in important respects, an *analogue* of the latter. The pre-scientific cultural condition, as illustrated by primitive geostatic astronomies, recapitulates on a cultural level certain basic features of the initial egocentrism of the child. Correspondingly, movement away from this cultural starting point of science parallels the developmental process by which the child's egocentrism is transcended. I have pointed out that individual cognitive development involves moving away both from reliance on immediate crude observation, and from uncritical acceptance of imaginative speculation not based on observation. Each of these developmental tendencies has a cultural counterpart in the process of scientific development.

In science, as in individual cognitive development, conceptual structures are formed which remove problems from the realm of immediate crude observation. Such structures include hypothetical premises linked to observational data via logic. This means, first, that logic comes to be used to extend the range of implications of observed facts. A young child, just learning to count, when asked how many apples he would have if he began with two and acquired two more, might want to perform the experiment and count, while an older child or adult, at least in our culture, would be more likely to recognize the problem as a strictly logical one requiring no observation. Science continues, on a cultural level, this process of removing from the observational realm those problems and aspects of problems for which logic alone suffices. But logic does more than merely broaden the range of conclusions that can be drawn from observational data. It also becomes integrated with observation, in a way which transforms and extends the observational process itself. Specifically, it is used to organize observations in ways which increase their capacity to yield generalized information, as in the construction of controlled experiments. Inhelder and Piaget report that the European adolescents whom they studied have characteristically acquired an implicit understanding of the significance of controlled experimentation.[5] At the cultural level, the systematic exploitation of this procedure is commonly recognized as one of the central features of science.

In science, as in individual cognitive development, movement away from reliance on crude observation is paralleled by movement away from uncritical acceptance of non-observational impressions and ideas. Such movement is reflected, at the individual level, in the contrast between the child who makes no distinction between "perception" and "imagination" and the adult who makes this distinction in principle even if he makes some errors in applying it. At the cultural level, we may note a comparable contrast between the pre-modern confusion and the modern separation of scientific concepts and speculative (including theological) concepts. Moderns who examine early scientific writings are likely to notice not only ideas which are now regarded as false, but also, mingled with reports of observational data, purely imaginative ideas which today would be considered not as false but as scientifically irrelevant, e.g., ideas about the role of God in movements of celestial bodies.

Each of the two aspects of scientific development discussed above has a familiar label. The first is commonly known as the "rational" aspect and the second as the "empirical" aspect. It has often been said that science involves an integration between "rational" and "empirical" tendencies. The analysis presented above suggests a way in which each of these two aspects of science may be identified as an analogue and extension of a corresponding aspect of individual cognitive development.

Both the "rational" and the "empirical" aspects of cognitive development have been described above primarily in negative terms: i.e., each has been described as a process of moving away from certain initial conditions. This description may now be formulated in a more positive way. Corrections for observational limitations cannot be achieved either by declining to make inferences or generalizations beyond what has actually been observed, or at the other extreme by drawing speculative conclusions which are completely arbitrary in relation to observed facts. What is needed, if the correction process is to occur, is something intermediate between these extremes: a type of conceptualization which goes beyond actual observations, suggesting ways in which observations are to be interpreted, yet which is also subject to correction on the basis of subsequent observation. The process is thus circular: conceptualizations "correct for" observational limitations, and may themselves be corrected by subsequent observation in turn.

Such linking of conceptualization with observation produces a developmental process which is continuous in some respects and discontinuous in others. There is continuity in the progress which is made toward greater observational knowledge and toward more systematic ordering of such knowledge. There is also discontinuity with respect to particular conceptualizations: new concepts and propositions are not merely superimposed upon older ones, but rather displace the latter to some extent. When an episode of transition from one conceptual system to another is viewed from the perspective of the older system which is displaced, the *discontinuous* aspect is likely to appear particularly prominent, so much so that the continuity in direction of development may be obscured from view. Thus, when a transition is made from one scientific system to another, there are commonly some adherents of the old system who regard its displacement as a break with scientific progress rather than as a continuation of it. On the other hand, when cognitive development is examined retrospectively, there seems to be a tendency for the continuities to be noticed and remembered, and for the discontinuities to be over-looked or minimized. The amnesia of the older child or adult, who cannot remember having experienced conceptual transformations in his own intellectual development, and who regards his own past learning as simply a matter of quantitative accumulation of knowledge,[6] is paralleled by the "invisibility of scientific revolutions" which has led scientists to overlook the conceptual transformations in the histories of their own disciplines, and to interpret these histories exclusively in terms of a quantitative accumulation model.[7]

Science as an Outgrowth of Traditional Cultural Knowledge

In the preceding section, science has been compared with a process which is similar to it in being cognitive and developmental but different in being individual rather than cultural. Now we may compare it with another phenomenon which is similar to it in being cultural and cognitive but different in that it lacks the intrinsically developmental character of science. Specifically, science will be compared with traditional, pre-scientific cultural knowledge or belief systems. (No distinction will be made between "knowledge" and "belief" here.)

I shall begin by discussing some general characteristics of cultural knowledge systems, characteristics which are found both in primitive cultures and in the culture of modern science, and shall then turn to the distinctive features of science as an outgrowth of pre-scientific cultural knowledge.

A "cultural knowledge system," as the term is used here, is any set of ideas, prevailing in a given culture or subculture, which provides a way of organizing information about the world or about any aspect of it. Cultural knowledge systems in this sense include "world views," philosophies, theologies, political ideologies, and scientific theories, as long as such systems prevail in some cultural context. Cultural knowledge systems may be identified at different levels of generality; thus Christianity is one such system, and each of the various competing theologies which offers a distinctive interpretation of Christianity is also such a system.

A cultural knowledge system does not contain a random assortment of propositions. Rather, there is a tendency for the propositions of such a system to be constrained or patterned in certain ways. Partly, this is a matter of definition: if ideas were accepted randomly, they could hardly be said to constitute "knowledge systems." But it is also an observable reality that randomized collections of propositions are characteristically not culturally established, while knowledge systems, which are non-randomly patterned, are found in all cultures.

We may recognize two different kinds of constraints on the range of propositions included within a cultural knowledge system. There are *rational* constraints, pertaining to the internal organization of the system (i.e., to its *systematic* character), and *empirical* constraints, pertaining to the system's relation to observed facts (i.e., to its character as a *knowledge* system). Rational constraints involve tendencies to develop and maintain *internal coherence* — logical consistency and mutual reinforcement among the propositions of the system. Empirical constraints involve tendencies to develop and maintain *factual plausibility* — compatibility with accepted "facts," and a capacity to provide satisfying explanations of those accepted facts which the system is expected to explain.

The idea that cultural knowledge systems tend generally to be constrained in both rational and empirical respects may be misunderstood, and rejected on the basis of misunderstanding, unless allowance is made for the full diversity of ways in which such constraint may be manifested.

Rational constraints may involve internal changes within a system (e.g., elimination of a proposition which contradicts other propositions of the system), or may involve the differentiation of a system into competing variant systems (e.g., the proliferation of variant doctrinal versions of Christianity, Marxism, the biological theory of evolution, and psychoanalytic theory). Such diversification may, of course, increase the amount of doctrinal conflict within the culture as a whole, yet it may nevertheless constitute a rationalizing movement in the present sense if each of the emerging doctrinal variants is more internally coherent than the original parent doctrine. I am not claiming, as

Ruth Benedict claimed,[8] that entire societal cultures tend to become "integrated," but rather only that such an integrating tendency applies to each separate knowledge system within a societal culture, and this tendency may sometimes produce what appears to be the opposite of "integration" from the standpoint of the culture of society as a whole.

Empirical constraints may involve changing a system to bring it into line with observed facts, or reinterpreting the facts to fit them into the system. The latter, although it may sometimes produce a distortion of reality, may also sometimes have exactly the opposite effect, and may even play not only a legitimate role but a crucial one in the scientific process. When, for example, the movements of the planet Uranus failed to conform with predictions made on the basis of the Newtonian system, that system was *not* abandoned or modified to bring it in line with the new and unexpected facts. Instead, the unpredicted deviations of Uranus were reinterpreted by assuming that the basic Newtonian premises were valid, and that a hitherto unknown planet was exerting a gravitational influence which had not been taken into account in the original calculations. The outcome of this attempt to fit anomalous observations into an established system was the discovery of Neptune[9] — a discovery which could hardly have occurred if the Newtonian system had been hastily abandoned or changed on the ground that it was no longer consistent with observed facts.

I have identified various mechanisms through which rational and empirical constraints may be effected, and have suggested that, when all of these mechanisms are allowed for rather than merely certain ones in particular, rational and empirical constraints appear characteristic of cultural knowledge systems generally. There is one other distinction which may be made, between mechanisms of constraint, which cuts across the preceding distinctions. Rational and empirical constraints may emerge "naturally" through protracted cultural-evolutionary processes, or "artificially" through deliberate construction by particular individuals. Hence, paradoxically, we find these constraints most perfectly displayed in two drastically different types of situations. On the one hand, a high level of internal coherence and plausibility often characterizes traditional knowledge systems which have matured over a long period of time in relatively static and isolated cultures, under conditions which have permitted a gradual, spontaneous "weeding out" of imperfections. Thus, some primitive witchcraft and magical systems have been described as composed of ideas which are neatly interlocked in self-reinforcing patterns and as admirably adapted to the task of providing satisfying explanations of relevant observed facts, e.g., explanations of why one person becomes ill while another does not.[10] On the other hand, some cultural knowledge systems, including those of science, are deliberately constructed to possess internal coherence and plausibility and, in the case of scientific systems, to possess certain other features also, which will be discussed later.

I am suggesting here that some cultural knowledge systems of primitive peoples, including some witchcraft and magical systems, are quite similar to the cultural knowledge systems of modern science, in terms of certain general structural features: similar in that they possess "elegant" internal organizational patterns, and similar in their superbly effective adaptation to whatever factual realities may be recognized in the given context. Of course these two kinds of system are drastically different in some other respects, but the differences can best be viewed within the framework provided by a preliminary outline of similarities.

These similarities are especially noticeable if the systems in question are contrasted with others which have emerged without a long background of cultural evolution under stable conditions and also without deliberate theoretical construction. For example, ideas underlying ethnic prejudice in the United States are often manifestly inconsistent: the same person who insists that Negroes are "satisfied" with their status might also accuse them of making "unreasonable demands," and the same person who criticizes Jews for being "unwilling to assimilate" may also complain that they "hide" their Jewish characteristics.[11] Such inconsistencies indicate cultural knowledge systems in which rational and empirical constraints are relatively weak, both in comparison with many traditional knowledge systems prevailing in highly integrated primitive cultures, and in comparison with many deliberately constructed modern knowledge systems including those of science.

In discussing cultural knowledge systems I have distinguished between features of the internal organization of such a system, and features of the relation between the system and observed facts external to it. In defining the concept of *scientific system* it is possible to make a similar distinction, in a way which will be reasonably consistent with familiar terminological usage and which also, at the same time, will enable us to analyze scientific systems within the same frame of reference already presented for the analysis of cultural knowledge systems in general. Specifically, we may recognize a system of ideas as "scientific" to the extent that it combines two features: *abstractness*, which pertains to the internal organization of the system, and *testability*, which pertains to the system's relation to external facts. Abstractness means that the logical aspects of the system's internal organization are differentiated from the factual content of the system: that logical relationships among the component propositions are clarified, and that the most general propositions of the system pertain to principles or laws to be applied under hypothetical circumstances, rather than to concrete factual situations. *Testability* means that the system provides a basis for predicting observable outcomes, and hence that it is subject to evaluation in terms of the accuracy of its predictions.

Either abstractness or testability may be present without the other. Abstractness may be found in various philosophical and theological systems which are not testable. On the other hand, it is possible to have highly developed bodies of testable knowledge which lack abstractness; for example, repeated observations of sunrise, sunset, lunar phases, stellar movements, and other astronomical phenomena may provide a reasonable basis for predicting such phenomena through simple, concrete projections of past trends into the future, without any abstract formulations of astronomical "laws." In fact, abstractness and testability are relatively difficult to combine, because abstractness means that a system, if it is to be testable, can be tested only under special conditions which must usually be artificially constructed. Specifically, the combination of abstractness and testability, which marks a system as "scientific," means that the system predicts certain observable outcomes rather than others, not under ordinary or natural conditions, but under ideal conditions of observational or experimental control, e.g., "in a perfect vacuum," "without friction," "with impurities eliminated," "with both groups equated in specified respects."

This definition of a scientific system makes no reference to the *content* of the system, i.e., to the actual ideas which it embodies.[12] Instead, the definition is formulated strictly in terms of structural features. It follows that a system may be "scientific" in the present sense, even if its content is of a sort which is generally associated with non-scientific cultures, as long as the essential structural features are present. We may illustrate this point, and in doing so also illustrate the meaning of scientific system as here defined, by constructing a hypothetical system involving ideas of witchcraft and magic, which nevertheless meets the stated structural requirements for a scientific system. Let us suppose that, in a given society, people generally believe that (a) witchcraft tends to produce illness in intended victims, (b) illness is produced only by witchcraft, (c) magical defenses tend to protect a person from (witchcraft-induced) illness, and (d) illness in an intended witchcraft victim can be warded off only via magical defenses. We have here a system which as described is "abstract" in that it tells us what will happen under certain conditions of perfect experimental control: victims of witchcraft who are not protected by magic will fall ill. The system is also *testable* in principle: we could attempt to make someone ill via witchcraft, prevent the use of defensive magic, and see if he falls ill as the system would predict under these conditions.

Since witchcraft and magical systems as we find them in "real life" are characteristically *not* "scientific," we may ask precisely how they differ from this artificially constructed system. An answer to this question may help to clarify further the meaning of a scientific system. Appropriate answers would presumably include the following points, and perhaps other points also:

1) The propositions of actual witchcraft and magical systems are ordinarily stated only in concrete terms, with reference to particular situations; e.g., "X used witchcraft to make Y ill." Abstract propositions of the "if . . . then . . ." type — e.g., "if witchcraft and no defensive magic, then illness" — are ordinarily not formulated by believers in such a system, even though the external observer from a scientifically advanced culture may recognize such abstract ideas as implicitly underlying the concrete statements and the behavior of the believers.

2) An actual witchcraft and magical knowledge system is also likely to be untestable in practice, by virtue of unclear criteria for determining when witchcraft and magic are being employed. When a person becomes ill without any manifest use of witchcraft the possibility that witchcraft was employed cannot ordinarily be ruled out, because witchcraft may be undertaken secretly. When witchcraft is manifestly employed and the intended victim does not fall ill, the efficacy of witchcraft is not seriousy challenged because the failure can be interpreted as a consequence of secret defensive magic, or a consequence of unobserved errors in the witchcraft process itself.

In such ways as these, traditional cultural knowledge systems generally fail to meet the requirements for "scientific" status. And, even systems which are essentially scientific may "degenerate" in this respect, or may be applied in ways which give them a non-scientific status in particular situations. To illustrate these latter possibilities, and to illustrate also the delicate balance which the combination of abstractness and testability entails, let us consider again the previously-mentioned problem of relating the orbit of Uranus to the Newtonian system. Application of that system to the task of predicting the movements of Uranus in the early nineteenth century led to predictions which diverged from the actual observed path of the planet. Let us suppose that two astronomers, confronted with this discrepancy, respond in different ways: one insisting that the Newtonian system has been disproven by the discrepancy and must be abandoned, the other insisting that the Newtonian system is permanently valid and that the discrepancy is to be explained in terms of an unknown planet influencing the movement of Uranus, or in some other way which preserves the Newtonian system intact. If each insists that his point of view is unquestionably true, rather than merely a possibility to be considered, then each will be making an error of a different kind: one will be denying or negating the *abstract* character of the Newtonian system, while the other will be denying or negating its *testable* character. The error of the first scientist consists in failing to allow for the distinction between the Newtonian system itself, and one particular interpretation of the system. By assuming that the system itself has been disproven, merely because one interpretation of the system which makes no allowance for unknown planets has been found faulty, this scientist is ignoring the possibility of considering the Newtonian system *in abstraction from* any particular interpretation of it. On the other hand, if the second scientist insists that the Newtonian system must be considered valid and that some satisfactory way will always be found to resolve any apparent discrepancies between the system and "the facts," he is denying the possibility that the system can ever be effectively *tested* in terms of empirical evidence. To preserve the combination of abstractness and testability which marks a system as "scientific," one must allow *both* for the possibility that an apparent discrepancy between the system and observed facts may be eliminated by a reinterpretation of the situation in which the observations were made, and for the possibility that such a discrepancy may sometimes be so serious and so persistent, despite attempts to deal with it via "reinterpretation," that abandonment of the system itself is appropriate. Unfortunately from the standpoint of clarity in scientific procedure, there does not exist any reasonably precise, objective and general criterion for predicting or deciding which of these two possible outcomes is relevant in any particular case.

As the preceding discussion suggests, the structural requirements for a scientific system are difficult to achieve and maintain. It is similarly difficult for a scientific system to achieve and maintain cultural acceptance within the scientific community, and hence to become and remain a "cultural knowledge system." Many systems which are genuinely scientific in terms of the structural features of abstractness and testability have nevertheless been displaced by other scientific systems, and are thus of historical interest only, or else have never been accepted at all by the scientific community (or by any other group, although this latter hypothetical possibility is actually trivial, since the scientific combination of abstractness and testability has not been effectively maintained except in the specialized context of the scientific community). In other words, just as most cultural knowledge systems are not "scientific," so likewise many scientific systems are not culturally accepted and hence are not "cultural knowledge systems." And, just as we previously asked what makes a cultural knowledge system "scientific," so likewise we may now ask what makes a scientific system culturally accepted, and hence a "cultural knowledge system," within the framework of the scientific community.

"Internal coherence" and "plausibility" are necessary for acceptance of a scientific system within the scientific community, just as these attributes are necessary for the acceptance of any knowledge system in society generally. However, there are additional requirements as well, for scientific acceptability. If a system is abstract, it becomes meaningful to ask about the extent to which it is characterized by "simplicity." In other words, abstractness provides a context within which simplicity becomes a meaningful criterion for the evaluation of a system. Similarly, if a system is testable, it becomes meaningful to ask about the extent to which it passes the relevant tests, i.e., the extent to which it provides the capacity to predict accurately rather than merely to formulate plausible retrospective explanations. The criteria of *simplicity* and *predictive capacity* which thus become relevant to the evaluation of scientific systems have two features which have important implications for the scientific process. First, these criteria provide a basis for the *comparative* evaluation of *competing* systems in terms of their relative acceptability under prevailing conditions. Secondly, these criteria provide no basis for assuming that any given system is permanently superior to all potential competitors in relevant respects. Hence, scientific systems have, at best, a permanently tentative acceptability.

In the preceding section, science was viewed as a cultural extension and counterpart of individual cognitive development. Here, I have presented another view of science, as a developmental extension of traditional cultural knowledge systems. Such traditional systems have sometimes achieved high levels of internal coherence, with each component part of the system elegantly integrated into the system as a whole, and also a high level of plausibility, with all facts which the system is "supposed to" explain, actually explained in a satisfying way within its framework. Science does not necessarily improve upon such performances of traditional knowledge systems, but rather introduces new criteria of performance, new criteria which, however, are merely specialized extensions of the old ones. Science involves systems which go beyond the achievements of traditional cultural knowledge systems, in the directions of abstractness and testability, and it involves a sequence of displacements among such systems, in terms of the criteria of simplicity and predictive capacity.[13]

Science as a Cognitive Form of Cultural Development

Biological evolution has proceeded very slowly by comparison with cultural development, or cultural "evolution" as it is often called, in those situations in which the two can be meaningfully compared with respect to speed. Thus, men of today are evidently so little changed, biologically, from their ancestors of about 40,000 years ago, that they are considered as belonging to the same species and subspecies. In particular, there is no evidence of improvement in innate mental capacity during this time. On the other hand, during this same time period, cultural development has produced one drastic transformation after another in the conditions of human life, and produced such an immense and rapidly-expanding accumulation of knowledge that ordinary school children can know more about many aspects of the world around them than the greatest thinkers of only a short time ago could have possibly known.

Cultural development was remarkably rapid in comparison with biological evolution even in pre-scientific times. But since the rise of modern science in the seventeenth century, the pace of cultural development has been drastically accelerated, with progress in science and technology largely preceding and stimulating developmental changes in other aspects of culture. Thus, while cultural development is a relatively rapid "substitute" for biological evolution as an agent of change in human life, science in turn may be recognized as a "speeded up" form of cultural development, with a specialized emphasis on the cognitive aspect of culture, and in particular on cognitions related to the natural world.

As a basis for analyzing science within the framework of this perspective, we may briefly identify the mechanisms through which the process of biological evolution takes place. This process involves three distinguishable phases: 1) genetically determined variation among individuals of a given species, 2) consistent natural selection, with certain kinds of variants being more likely than others to survive and/or reproduce, and 3) hereditary transmission of the characteristics of the positively selected variants to their offspring. Thus, the evolutionary development of large species of horses from small ones, over a long period of time, may be explained very roughly as follows: 1) in each of numerous generations of horses, there was genetically-determined variation in size, 2) larger horses, for some reason which need not concern us here, were better able to survive and/or reproduce than smaller ones, and 3) genetically-determined tendencies toward "largeness" thus came to be increasingly prevalent over a number of generations.

The relevant evolutionary mechanisms may also be described more abstractly, without specific reference to biological concepts such as "species," "reproduction," and "heredity," and therefore in a way which provides a more widely applicable model of the evolutionary process. Donald T. Campbell, for example, has described evolution as involving 1) variations, 2) "consistent selection criteria," and 3) "the preservation, duplication or propagation of the positively selected variants."[14] This model can be applied to cultural as well as biological phenomena. In such an application, one would assume 1) variation among cultural objects of a given type, 2) selective survival and diffusion of some of the variants, and 3) retention of the positively selected variants as part of the established culture, as a starting point for subsequent variation.

Processes of cultural development differ from the process of biological evolution in man and closely related species, in several ways, including the following:

1) The long interval of time between birth and procreation limits the speed with which relevant variations may be introduced into the biological evolutionary process. Cultural development, on the other hand, has no dependence on comparatively slow mechanisms of biological maturation and reproduction; it takes place instead through learning, which is not similarly limited in its potential speed.

2) Complex biological organisms, including the human organism, require a high level of integration among different aspects of their functioning. This means that a change may not be viable unless a number of other changes occur at the same time. This requirement of internal integration is a powerful stabilizing force which discourages rapid evolutionary change at the biological level. Major variations suddenly introduced through mutations are generally fatal, because the necessary associated changes do not occur at the same time. Cultural systems, on the other hand, require comparatively little integration. It is thus relatively easy for change to occur in one aspect of a culture, with other aspects left relatively intact at least for a short time, without the survival of the entire culture thereby being threatened.

3) In a "perfect" evolutionary process, the initial variations are random with respect to the direction of evolutionary development, and this direction is determined by consistent selection criteria. Biological evolution and cultural development tend to deviate from this pattern in opposite ways:

a) Biological evolution is affected by considerable randomness not only in the initial variations which provide raw material for natural selection, but also in that selection process itself. The individual organism is exposed to environmental hazards of such variety that the advantages conferred even by those variations which lie in the direction of evolutionary development may be only very slight. Thus, even if greater intelligence makes it easier for an organism to survive and reproduce, that trait might make survival only very slightly easier, so that many comparatively "stupid" individuals will survive and many comparatively intelligent ones will fail to do so; the advantages conferred by greater intelligence will thus have an evolutionary effect only over many generations.

b) In cultural development, there is also some randomness in selection. However, foresight and planning tend to produce relatively great consistency, not only in the selection process, but also in the initial variations: innovators tend to develop new ideas which they believe are likely to be accepted, rather than to let ideas emerge "randomly." There is, accordingly, a tendency for variations to anticipate the criteria of selection which will be applied to them, and to depart from randomness in the direction indicated by these criteria.[15]

Each of the differences noted above, between biological evolution and cultural development, gives the latter a potential advantage with respect to speed. In science, these potential advantages are most fully manifested. Science can be interpreted as a process of cultural evolution, in which competing scientific propositions and systems provide the relevant variations, with simplicity and predictive capacity as the selective criteria, and with patterns of communication and socialization within the scientific community making possible the retention of those propositions and systems which are positively selected. Science is, furthermore, a process in which the evolutionary mechanisms operate with impressive efficiency.

It is true that scientifically irrelevant considerations often interfere with the perfect application of scientific selective criteria; that scientists often accept or reject proposed innovations for reasons not entirely "scientific." Such departures from scientific standards will appear quite common if we compare what actually happens in this respect with what "should ideally" happen. On the other hand, departures from perfect consistency in application of scientifically relevant selective criteria appear remarkably *in*frequent and *un*important when comparison is made not with a hypothetical ideal situation which has never existed in real life, but rather with similar departures from perfect consistency in pre-scientific cultural-evolutionary processes and in biological evolution. Perhaps it would be appropriate to say that science is, in this respect (although not in all other respects), analogous to the kind of artificial biological evolution which is employed in the deliberate breeding of new forms of domesticated animals and plants, in which criteria of selection are highly consistent because the process of selection is deliberately controlled.

Science as a Cultural Process

The superbly efficient system of selection in science is supported by equally efficient mechanisms of variation and retention. As science progresses, scientists learn from experience that certain kinds of innovations are more likely to be favorably selected than others and accordingly tend to avoid formulations which manifestly fail to meet prevailing scientific standards. Within the limitations imposed by these standards, there is strong encouragement to develop potentially acceptable innovations, and efficient retention of favorably selected innovations is provided for by organizational arrangements within the scientific community.

Science thus appears as a rapid process of cultural development in contrast to cultural changes which, as in the case of fashions, may be rapid but have no clear consistency in direction and are thus not developmental. It contrasts also to cultural development processes such as that of "religious evolution," in which a consistent direction of development may be identified through historical investigation by scholars,[16] but with less efficient evolutionary mechanisms and hence slower and more irregular developmental progress. This feature of science reflects its status as a process maintained within a specialized and relatively autonomous community which is organized to facilitate consistently directed change.

Summary

Science has been identified as a process which is cultural, cognitive and developmental. Each of the three aspects of this conception provides a basis for a different view of science.

The three resulting views of science represent not three different phenomena, or three opposing theories, but rather a single phenomenon seen from three different perspectives in terms of three different comparisons. These views are mutually compatible, and when taken together they provide a conception of science richer than any one or two of them alone.

According to this conception, science is a process of transition in which cultural knowledge systems with certain characteristics are displaced by other knowledge systems sharing these same characteristics but which also differ in certain respects from those which they displace. The characteristics shared both by the displaced systems and by their replacements are "abstractness" and "testability:" these are criteria by which a system is admitted into the arena in which it will compete with alternative systems sharing the same features, for acceptability in the prevailing body of scientific knowledge. The systems which compete successfully at any given time, thus displacing whatever alternative systems may have been prevailing, do so by virtue of superiority in terms of the criteria of "simplicity" and "predictive capacity." The mechanism by which the scientific process occurs involves selection, on the basis of "simplicity" and "predictive capacity," from among variant alternative systems, all of which share the features of abstractness and testability which are essential for entrance into the competition for selection.

The *direction* of scientific development is similar to that of individual cognitive development. The *starting point* of scientific development is traditional cultural knowledge. The *structure* of scientific development is similar to that of the evolutionary process in general and the process of cultural evolution in particular. Science is an extension of cognitive development from the individual to the cultural level, and a developmental outgrowth of traditional cultural knowledge, and a specialized cognitive variant and extension of cultural evolution.

This conception does not reflect science as the scientist himself is likely to see it in the course of his own work. The purpose has been to go beyond that perspective, to define science in terms of what seem to be the most central tendencies in the most highly developed scientific disciplines over long periods of time. Many scientists do not consider their task to be the displacement of an established system in their area of specialization, but this nevertheless happens in science in the long run, and when an episode of this sort is examined retrospectively, it commonly appears that some scientists contributed to it without intending to do so and without realizing that they were doing so. A scientist may make a genuine contribution to science in a way which strengthens an already established system, e.g., by demonstrating its applicability to a new class of phenomena, but in doing so he may very well be contributing also to the ultimate displacement of that system by opening up the possibility of further inquiries which will expose serious defects in the system.

The conception of science presented here diverges not only from what scientists themselves are likely to consider science to be, but also from the most familiar sociological approaches. It is based on analogies with phenomena outside the sociological realm (e.g., individual cognitive development and biological evolution), and makes no use of sociological concepts such as "institution," "occupation," and "profession" which suggest or imply a relatively stable relationship between science and society. This, however, seems entirely appropriate, for it merely reflects ways in which science itself differs from many other phenomena studied by sociologists.

NOTES

1. Discussions of the analogy between science and individual cognition, from varying points of view, will be found in Arthur Koestler, *The Sleep Walkers*, London: Hutchinson Press, 1959; Donald T. Campbell, "Methodological Suggestions from a Comparative Psychology of Knowledge Processes," *Inquiry*, 53 (Autumn, 1959), pp. 152-182; and various works by Jean Piaget, which are nicely summarized in J. H. Flavell, *The Developmental Psychology of Jean Piaget*, Princeton: Van Nostrand, 1963.

2. The concept of egocentrism is discussed in various works by Jean Piaget, including *The Construction of Reality in the Child*, New York: Basic Books, 1954. See also Flavell, *op. cit.*

3. For interesting illustrations of this point, see Max Wertheimer, *Productive Thinking*, New York: Harper, 1945, and Leonard Schatz-

man and Anselm Strauss, "Social Class and Modes of Communication," *American Journal of Sociology*, 60 (January, 1955), pp. 329-338. Wertheimer discusses individual variation in the capacity or tendency to integrate diverse perspectives, while Schatzman and Strauss discuss social class variations.

4. Quoted by Stephen F. Mason, *A History of the Sciences*, New York: Collier, 1962, p. 121.

5. Barbel Inhelder and Jean Piaget, *The Growth of Logical Thinking from Childhood to Adolescence*, New York: Basic Books, 1958.

6. Anselm Strauss had described such amnesia as "a kind of systematic forgetting that has nothing to do with repression." "The Development and Transformation of Monetary Meanings in the Child," *American Sociological Review*, 17 (June 1952), pp. 225-286.

7. Thomas S. Kuhn, *The Structure of Scientific Revolutions*, Chicago: University of Chicago Press, 1962.

8. *Patterns of Culture*, Boston: Houghton Mifflin, 1934.

9. The history of this episode is described in Herbert Hall Turner, *Astronomical Discovery*, Berkeley, University of California Press, 1963.

10. Relevant features of magical and witchcraft systems are discussed by E. E. Evans-Pritchard, *Witchcraft, Oracles and Magic Among the Azande*, Oxford: Clarendon Press, 1937; Max Gluckman, *Custom and Conflict in Africa*, Glencoe, Ill.: Free Press, 1955; chap. 4, "The Logic in Witchcraft," pp. 81-108; and Michael Polanyi, *Personal Knowledge*, London; Routledge and Kegan Paul, 1958, 287-294.

11. Such inconsistencies have been analyzed by John Dollard, *Caste and Class in a Southern Town*, New Haven: Yale University Press, 1937; Gunnar Myrdal, *An American Dilemma*, New York: Harper, 1944; Gordon W. Allport, *The Nature of Prejudice*, Cambridge, Mass.: Addison-Wesley, 1954, and Robert K. Merton, *Social Theory and Social Structure*, Glencoe, Ill.: Free Press, 1957, chap. 11, "The Self-Fulfilling Prophecy," pp. 421-436.

12. I have been using the term "scientific knowledge" to refer to knowledge acquired through the scientific process and accepted by scientists generally at some point in the history of science, regardless

of the extent to which such knowledge is organized in "systems." The term "scientific system" is defined here in a quite different way, to refer to any system with certain structural features, regardless of whether such a system has ever been accepted by scientists. These two concepts should not be confused.

13. These two criteria represent partially conflicting tendencies which must be mutually adjusted. Efforts are made to increase simplicity without interfering with predictive capacity, to increase predictive capacity without reducing simplicity, and to find a satisfactory balance between the two criteria when no way is found to improve the status of a system in one of these respects without a corresponding deterioration in the other respect. Rules for making decisions in situations of this latter type are only poorly developed and, in any case, need not concern us here. It is relevant here, however, to note that compromises are often necessary. For a discussion of this problem, see Peter Caws, *The Philosophy of Science*, Princeton: D. Van Nostrand, 1965, chapter 31, "The Choice Between Alternative Theories," pp. 232-239.

14. Donald T. Campbell, "Variation and Selective Retention in Sociocultural Evolution," in Herbert R. Barringer, George I. Blanksten, and Raymond V. Mack, *Social Change in Developing Areas*, Cambridge, Mass.: Schenkman, 1965, 19-49, p. 27.

15. A tendency of this sort has also been described in biological evolution, (see Lancelot Law Whyte, *Internal Factors in Evolution*, New York: George Braziller, 1965). If we assume that there is such a tendency at the biological level, then we arrive at the following formulation. A "perfect" evolutionary process entails random variation and consistent selection. At both the biological and the cultural levels there is divergence from "perfect" evolution in two different ways: there is some randomness in the selection process, which interferes with perfect consistency in selection, and there is some consistency in the initial variation, which represents a departure from perfect randomness. The first form of divergence is most prominent at the biological level and the second at the cultural level.

16. Robert N. Bellah, "Religious Evolution," *American Sociological Review*, v. 29, June, 1964, pp. 358-374.

Chapter 4

ASPECTS OF THE DEVELOPMENT OF SCIENCE

"Science" and "The Development of Science"

The preceding chapter has discussed the *nature* of science. This chapter will discuss the *development* of science. A preliminary comment about the distinction between these two topics may be appropriate.

Science has been defined as a certain process of development. The distinction, then, is between the developmental process which *is* science, and the development *of* this developmental process. The developmental process which *is* science is a process of a certain general type, which can be (and has been) formulated in abstract terms. By contrast, the development *of* science was a unique, although protracted, historical event, which is relatively difficult to analyze or explain because its uniqueness precludes the kind of comparative inquiry which would provide the most effective basis for analysis or explanation.

The history of science cannot be divided sharply into two eras, one involving the development of science and the other involving the continued occurrence of the scientific process after its development. These two phenomena have not been totally separated; rather, some historical events have involved mixtures of the two. Thus, the transitions from the physics of Aristotle to the physics of Newton, and from the latter to the physics of Einstein, were partly transitions from "pre-science" to "science," but also partly transitions *within* science, from one scientific or partly scientific system to another. A Newtonian physicist looking back on the Aristotelian system, and a contemporary physicist looking back on the Newtonian system, would each have good reason to conclude that the system he looked back on was in some respects scientific although less adequate than the system which replaced it, and in some other respects not genuinely scientific at all. Nevertheless, we can make a reasonable distinction between an historical period during which the development of science was largely, although not entirely, completed, and a later period during which further perfections of the scientific process, although important, were merely incidental by-products of the continued and expanded functioning of science as previously developed. The work of Sir Isaac Newton (1642-1727) represents an approximate boundary line between these periods. Thus, the shift from Aristotelian to Newtonian physics may be viewed primarily as a shift from "pre-science" to "science," and only secondarily and in limited aspects as a shift from one scientific or partly scientific system to another, while the later shift from Newton's system to Einstein's was primarily part of the ongoing process of science itself, even though it also involved, secondarily, some clarifications and perfections of this process. The Aristotelian system is primarily pre-scientific, although partially anticipating some features of scientific systems, while the Newtonian system is essentially scientific even though outmoded, but with some features reminiscent of pre-scientific days.[1]

Problems in the Study of the Development of Science

We cannot exclude the possibility that science will be ultimately self-terminating, through destructive applications of the technology which it has made possible, or even through substantial completion of the scientific process itself. However, science as we know it today has been self-reinforcing in certain limited but important respects: (1) The questions which scientists ask arise largely within the context of scientific inquiry itself, so that science is no longer dependent on its societal environment for stimulation and guidance, even though dependent financially and in various other ways. (2) Answers to scientific questions merely raise new questions, in a sequence which does not seem likely to reach a termination point in the near future. (3) Previous scientific achievements have produced a situation which encourages continued support of science, even though the arrangements for this support are potentially unstable and the amount of support is much less than many scientists would like.

These self-reinforcing features help us to explain the persistence of science. However, they are not particularly helpful when we seek to explain the early stages in the initial emergence of modern science, which is a much more complicated task. We must assume that the self-reinforcing features of science were not initially functioning, and must look elsewhere for "causes" of the development of science.

In a search for such causes, the questions for which answers are sought should be clearly formulated. Some questions which people ask about the development of science are trivial in the sense that the answers to them are obvious and not usefully informative.

One such question is: "Why did science develop?" Given hundreds of distinct cultures, differing among themselves and subject to change in innumerable ways, and given thousands of years of cultural history, it would be plausible to expect some society somewhere at some historical period, to "accidentally" combine those characteristics which create or constitute science. In this sense, the rise of science may be explained simply in terms of random variation among the cultures of the world over a long period of time.

Another question with an obvious answer would be: "Why did science develop independently on a massive scale only once, in only one of the world's numerous cultures?" The answer is that a massive emergence of science in one culture inevitably changed the world in ways which precluded any subsequent *independent* development of large-scale science elsewhere. Science has spread widely around the world, but this is a matter of cultural diffusion, not independent development. Once a major science-bearing culture came into existence, all other cultures which might have acquired science independently underwent major transformations under the impact of the scientifically-advanced culture, and lost the cultural autonomy which a truly independent development of science would require.

We have, then, two questions concerning the rise of science which are trivial in the forms in which they are stated. By making minor changes in these questions, we can eliminate the triviality of each. Instead of asking "Why did science develop?" (in at least one society, sometime and somewhere) we may ask "Why did science develop in the modern West?" Instead of asking, "Why did science develop massively only once?" we may ask "Why did science on a large scale not develop anywhere prior to its emergence in the West?" Or, to combine the two questions into one: "Why did science develop when and where it did, rather than at some other time and place?"

These revisions transform our initially trivial questions into a substantial one which, however, is quite difficult to answer because we lack an adequate comparative framework. From a methodological standpoint, the ideal procedure would be 1) to classify cultures according to whether they have, or have not, developed science independently, 2) to compare science-producing cultures with others, using the information thus obtained to infer the causes of the development of science where it has occurred, and 3) to test the resulting inferences by predicting the conditions under which such development will occur in the future. However, this ideal procedure is rendered largely ineffective by one fact: that large-scale science has developed independently only once, in the culture of the West.

Of course we can still compare science-producing cultures with others, and seek to infer the causes of the development of science where it has occurred, even though the category of "science-producing cultures" has only one case subsumed under it. Specifically, we can ask what distinctive features of seventeenth-century western Europe, which were absent from other cultures, may have been responsible for the one historic instance of independently developed modern science. We can also make comparisons within the general framework of modern science: e.g., between different periods in scientific history, different scientific disciplines, and scientific developments in different nations, even though the phenomena thus compared are not truly independent of each other. Furthermore, we can compare the development of modern science with other episodes of cultural creativity in art, literature, religion, technology and social organization, in various parts of the world and various historical eras.

Nevertheless, no matter how valuable inquiries along these lines may be, they cannot fully compensate for our inability to compare the rise of science in European civilization with any other large-scale independent development of science. Since the rise of modern science is essentially an event (or sequence of events) which has taken place only once and in a single civilization, we have no adequate basis for distinguishing between essential features of the rise of science and features which represent local historical accidents. Interpretations of the rise of science will thus inevitably contain an arbitrary element. The best we can do in this respect is to make the unavoidably arbitrary aspects of our interpretations as explicit, systematic and reasonable as possible, and to avoid arbitrariness beyond the unavoidable minimum. The conception of science presented in the preceding chapter will be helpful in this connection.

A Preliminary Framework

Science has been identified in the preceding chapter as a process in which knowledge systems characterized by abstractness and testability are displaced by other systems which share these attributes but which are superior in terms of simplicity and predictive capacity. This conception suggests that the development of science involved two distinguishable phases: 1) the emergence of knowledge systems of the "scientific" type, characterized by abstractness and testability, and 2) the emergence of the developmental process in which new scientific systems displace older ones on the basis of the criteria of simplicity and predictive capacity.

Cutting across this distinction between phases is another distinction which follows from the present conception of science, between two aspects of its development: 1) the emergence of features pertaining to the rational qualities of scientific systems (abstractness and simplicity), and 2) the emergence of features pertaining to the empirical qualities of such systems (testability and predictive capacity).

The analysis which follows will be organized around these two distinctions. The purpose is not to give a complete explanation of the development of science, but merely to identify general processes which were involved in this development, and which an adequate explanation would have to take into account.

The Cultural Background: Rationalism and Empiricism

The preceding chapter has pointed out that cultural knowledge systems generally tend to have certain rational and certain empirical features, but also that in the most pre-scientific cultural knowledge systems these features may be quite limited in certain respects. The limitations in question may be stated more specifically as follows:

1) Under maximally pre-scientific conditions, thinking may be guided by logical norms, but these norms are implicit, imperfectly distinguished from norms concerning the substantive content of thought, and relatively simple. Arguments will be accepted or rejected without distinction as to whether logical or factual criteria are employed in accepting or rejecting them. Cultural patterns, under these conditions, may attain internal coherence, but only through *ad hoc* adjustments, not through

explicit systematic construction. Logic is used only to eliminate internal inconsistencies and to make relatively simple deductions, not to produce abstract systems.

2) A maximally pre-scientific cultural knowledge system may incorporate considerable empirical knowledge. The natural world as it appears to men everywhere reveals regularities which provide a basis for making confident generalizations concerning such mattters as alternations of day and night, seasonal changes, and stages in the human life cycle. However, under maximally pre-scientific conditions, norms pertaining to relationships of ideas to empirical evidence are imperfectly differentiated from other norms. For example, the confident generalization that "all men die" may be mingled with confident generalizations concerning life after death, without distinction between confidence born of repeated observation and confidence sustained by the impossibility of finding disconfirming evidence. Under these conditions empirical knowledge is acquired casually rather than through systematic and sustained inquiry.

The limitations thus noted, in the rational and empirical qualities of the most pre-scientific knowledge systems, may be divided into two general categories. 1) Both logic and observation are imperfectly differentiated from the matrices in which they occur: there is imperfect differentiation between the logical statuses of propositions and the contents of these propositions, and imperfect differentiation between observed facts and assumptions accepted on a non-observational (e.g., "speculative") basis. 2) Logic and observation are not only imperfectly differentiated from related phenomena, they are also both weakly developed. There are no complex deductive systems, and no emphases on observational inquiry into the natural world.

"Rationalism" and "empiricism" may be defined here as cultural orientations which encourage further development and differentiation of the minimal rational and empirical features noted above. Rationalism emphasizes the importance of logic, and empiricism the importance of observation, in connection with the acquisition and validation of knowledge. Rationalism means that deliberate efforts are made to organize propositions into abstract, internally coherent systems — e.g., to derive new propositions from those already accepted, to subsume accepted propositions under still more general ones, to identify and eliminate latent inconsistencies among propositions, and to clarify the logical relations among the propositions which remain. Empiricism means attention to observational detail, and evaluation of ideas on the basis of observational criteria.

Rationalism and empiricism, although different from each other, are mutually compatible, and in fact are combined together in the culture of science. Furthermore, they share some important features:

1) Both rationalism and empiricism are approaches to the acquisition and validation of knowledge. They both reflect an interest in the same questions concerning the nature of knowledge, even though they answer such questions in different (but not necessarily incompatible) ways.

2) The rational "tool" of logic and the empirical "tool" of observation are similar in that both provide a basis for agreement and understanding among men who differ in values and in other aspects of culture. Men of divergent cultural backgrounds may find it easier to reach agreement on simple points of logic and on immediate observational facts, than on their respective interpretations of more complex aspects of the natural and supernatural worlds.

3) Rationalism and empiricism are similar to each other in that each of these has a strongly *radical* implication, when "radicalism" is defined as rigid adherence to a principle.[2] Rationalism means, among other things, that as long as certain propositions are accepted, all other propositions which can be logically deduced from them must be accepted also, no matter how absurd they may seem. Empiricism means that the question of what propositions are to be accepted is to be answered by nature rather than by the arbitrary fiat of men, with men thus committing

themselves in advance to accept whatever answers nature may give to their questions. Thus, rationalism and empiricism are similarly radical in that each entails a rigid adherence to a certain principle (a different principle in each case), in defiance of traditional beliefs if necessary. Each entails a commitment to follow the road to truth wherever that road may lead, even though they emphasize different criteria for identifying that road.

Rationalism has appeared prominently in a number of aspects of Western culture over the centuries: in Aristotelian logic, Euclidian geometry, Roman (and modern) law, medieval Christian theology, and modern bureaucratic organization and modern science. Similarly, the empirical tradition in the West has had a long life of its own, reflected in the accumulated astronomical records of antiquity, the naturalistic art of Greece, in Aristotelian physics which was closely based on ordinary observations (as distinct from observations under controlled conditions), in the emphasis on applied technology in western Europe since early modern times, in the vast influx of new knowledge arising from the voyages of discovery, and most recently in modern science.

Western civilization has been generally characterized by radical tendencies to follow accepted ideas to their ultimate conclusions, and by contrasts between different radical ideas which have interacted to produce drastic instabilities. Manifestations of this basic cultural trait include not only the traditions of rationalism and empiricism which constituted the historical background of modern science, but also the historic tendency of Western religions to make radical and often conflicting claims to absolute and exclusive truth, and diverse tendencies to bring society under the total domination of particular institutions (e.g., the Church in medieval Europe, the free market system in "laissez-faire" economic theory, the one-party state in modern totalitarian dictatorships.) The emergence of modern science was thus consistent with underlying persistent features of Western culture, features which appear clearly only when this culture is viewed in a broad historical and comparative perspective.[3]

The Emergence of Scientific Systems

The preceding section began with an outline of essential features of a maximally pre-scientific situation, in which rationalism and empiricism had not yet appeared as cultural traditions. We may now outline essential features of a situation which is also pre-scientific, but with the rational and empirical "ingredients" of science available in the culture even though not yet integrated.

In this situation of incipient science, rational and empirical traditions are manifested in different areas of culture and in rigidly circumscribed contexts. Rationalism is manifested in the construction, explanation, and justification of established philosophical and theological systems, and rival systems which might share the same rational qualities are not tolerated. Empiricism is manifested in inquiries directed toward the solution of practical problems, and empirical inquiries are recognized as appropriate only for such a purpose.

Furthermore, each of the two traditions is manifested in a form which inhibits integration between them: rationalism in abstract systems which are non-testable or at least not clearly testable, and empiricism in observations under ordinary conditions rather than controlled conditions relevant to the testing of abstract systems.

We thus have four features of the cultural situation of incipient science. Two of these features pertain to the rational tradition, and two to the empirical. Two involve restictions on the contexts in which the traditions may be manifested, and two involve restrictions on the forms which they may assume. These features may be identified more specifically as follows: 1) *the requirement of doctrinal conformity*, which limits rationalizing to "acceptable" systems, and excludes manifestations of rationality on behalf of rival systems; 2) *the requirement of practical utility*, which limits empirical inquiry to recognizably "useful" contexts; 3) *the invulnerability of abstract systems* to empirical disconfirmation, by virtue of their non-testable character; and; 4) *the concreteness of empirical inquiry*, by virtue of its concentration on ordinary rather than controlled conditions of observation.

This cultural pattern reflects a social structure in which the rational and the empirical traditions are maintained by different, and unequal, groups or classes within society: the former by intellectuals, who have comparatively high status, and the latter by craftsmen whose status is comparatively low.

Progress toward integration of the two traditions did not initially require decisive abandonment of the four conditions identified above. On the contrary, various circumstances made possible considerable progress in this direction within the framework of these conditions, and their decisive abandonment was necessary only in the final stage of such integration. Circumstances which facilitated this development include the following:

1) Restrictions imposed by the requirements of doctrinal conformity and of practical usefulness, have not been identically defined in all times and places. Different cultures and political units have had different doctrines, and different definitions of practical problems. For example, a geostatic astronomy has been dogmatically insisted upon in some pre-modern contexts, but not in others, and numerous astronomical predictions which had great practical importance in ancient Babylonia, because of their relevance to astrology, were of no importance in some other ancient cultures. Such diversity has made it possible to do in one time and place what could not be done in another.

Diversity of a sort which facilitated the emergence of science in this and perhaps other ways was initially provided by the passing of proto-scientific knowledge through a sequence of different cultures: e.g., from Babylonia and Egypt to Greece, from the Greeks to the Arabs, and from the Arabs to the western Europeans, although the actual historical development was more complicated than this very brief outline suggests.[4] Later, considerable diversity appeared within the framework of the single general culture which gave birth to modern science, with implications as follows:

a) Liberation from the requirement of doctrinal conformity was facilitated by a diversity of theological and philosophical systems, which competed with each other because they sought support from the same people and because their uncompromising claims were mutually incompatible. As Karl Mannheim has emphasized, men who are confronted with competing doctrines, and have no manifestly adequate basis for selecting between or among them, are likely to become sceptical of all such doctrines, and come to think more freely than they would under conditions of greater cultural consistency.[5]

b) Liberation from the requirement of practical utility was facilitated by a diversity of potential practical applications. When a single body of knowledge came to be recognized as useful for a wide variety of practical purposes, the experts who developed and transmitted such knowledge were able to avoid domination by any one particular interest group, and tended to symbolize and support their autonomy through a rationalistic emphasis on knowledge "for its own sake." Such a development is suggested by Florian Znaniecki's description of the way in which the role of specialist in discovery of new knowledge has become differentiated from the role of specialist in the application of knowledge to practical affairs.[6]

Philosophical and theological diversity had a negative effect on the status of philosophy and theology in society, because the diverse systems competed with each other, challenging each other's claims. On the other hand, diverse practical applications of knowledge do not primarily "compete" with each other, but rather reinforce each other. Hence the effect of diversity was diametrically opposite in the rational and empirical realms: intellectual diversity reduced the importance of the rational tradition, while diverse applications of practical knowledge increased the importance of the empirical tradition. Given the initial imbalance in the prestige of the two traditions, which reflected the relatively high status of intellectuals and the lower status of craftsmen, this change tended to promote an equalization of status. This, in turn, was conducive to the emergence of science, in which rational and empirical components are not separated by arbitrary assumptions to the effect that one is "more important" than the other.

2) Requirements of doctrinal conformity and practical utility were thus not primarily challenged in direct ways; rather, their effects were reduced by certain kinds of diversities. Similarly, the beginnings of movement toward "testability" in abstract systems, and toward observation under "controlled" conditions relevant to the testing of abstract systems, also occurred in somewhat accidental ways which did not initially involve direct and principled challenges to the prevailing limitations on rational and empirical traditions.

a) It is commonly recognized that theological systems are characteristically not "testable" in the sense in which scientific systems are. However, a theological system may come to be interpreted in a way which incorporates testable propositions, and may thus become exposed to the possibility of empirical challenge, even though this outcome is presumably not intended or anticipated by supporters of the system. Movement in this direction occurred in medieval Europe, where Christianity was interpreted to include various Aristotelian and other pre-modern assumptions about laws of physics and astronomy, assumptions which became so closely integrated with Christian religious beliefs that a challenge to them appeared as a challenge to Christianity itself.

This movement of Christianity in an "empirical" direction remained incomplete, and Christianity was able to survive and continue to flourish in the same society which produced modern science by disentangling itself from pre-modern assumptions about the natural world as these became scientifically discredited. The full integration between rational and empirical traditions which was ultimately achieved did not occur by transforming initially non-testable systems such as Christianity into fully testable ones, but rather involved the construction of entirely new systems, those of science. However, the earlier partial movement of Christianity toward a "testable" status was a matter of great importance in connection with the rise of modern science. It meant that scientific discoveries which might have been accepted rather casually in some other societies encountered relatively great resistance in Christian Europe, because these discoveries appeared to challenge the empirical foundations of the prevailing theology. However, it also meant that new scientific discoveries which otherwise would have been assimilated without major cultural changes and hence without a major scientific breakthrough, produced in Europe a gigantic cultural upheaval which cleared the way for a further massive expansion of science.

b) Through repeated observations and simple projection of past trends into the future, it is possible to make impressively accurate predictions of sunrise, sunset, lunar phases, positions of stars, planetary movements, eclipses, and other astronomical phenomena, without theoretical understanding of such phenomena. In fact, a tendency in this direction was characteristic of ancient Babylonian astronomy.[7] Such a development is possible only because in some important respects nature provides her own controls for astronomical observation. A feather may be blown by the wind and a rolling ball may be stopped by friction, and these effects must be discounted by experimental controls if we seek to discover laws of motion through observations of falling feathers and rolling balls. Celestial bodies, however, move through the heavens with manifestly greater regularity, in such a way that "laws" appearing to govern their motions may be discerned without similar deliberately-imposed controls.

On the other hand, some such "laws" appear to have occasional exceptions suggesting that the laws themselves may not be perfectly understood, and these exceptions, in turn, may ultimately appear to involve lawful regularities of their own. For example, solar eclipses appear first as irregular interruptions of normal predictable patterns, but recording of observations over a long period of time will show that eclipses occur at reasonably regular intervals. This kind of situation provides an ideal stimulus for the development of a conception of nature as operating according to hidden but discoverable laws, a conception which rejects both the view that nature is capricious and unpredictable in principle, and the view that nature is already fully understood on the basis of ordinary observation.

It is reasonable to suppose that methods of controlled experimentation have been used implicitly by intelligent people everywhere, in dealing with practical problems. However, the explicit formulation and systematic exploitation of this method, and its application to theoretical as distinct from practical problems, first emerged as a major cultural development in the early modern West in connection with the emergence of modern science. The fact that in certain respects observational situations were controlled by nature sufficiently to suggest the operation of incompletely revealed natural laws to men who observed carefully without imposing complicated, non-obvious controls of their own, gave an important impetus to the movement toward rational control of observation.

The obstacles to science which have been discussed here have not been fully overcome. In fact, modern totalitarian dictatorships tend to separate the rational and the empirical elements which science has united, forcing knowledge to be primarily responsive to ideological dogma rather than to empirical observation, and forcing empirical inquiry to develop in ways which are relevant primarily to immediate technological demands rather than to the evaluation of abstract systems.[8] Nevertheless, these obstacles have been overcome in various times and places in recent centuries in the modern West, to an extent sufficient to permit the spectacular growth of science which has occurred.

Movement toward convergence between rational and empirical traditions produced a decisive integration in the seventeenth century. Francis Bacon (1561-1626) and Rene Descartes (1596-1650) presented views of scientific method which showed a partial, but incomplete, integration. Bacon emphasized the importance of empirical observation and conceded a role, but not a sufficient role, for logic and mathematics, while Descartes emphasized the importance of logic and conceded a role, but not a sufficient role, for empirical observation.[9] The decisive breakthrough to rational-empirical integration came with the clarification by Galileo (1564-1642) of the methodology which links abstract hypothesizing with controlled experimentation, and with the formulation by Sir Isaac Newton (1642-1727) of a system which applied this methodology in such an impressive way that it came to serve as the fundamental model for scientific systems in general.

The Emergence of the Scientific Process

The emergence of scientific systems means that the range of acceptable questions is no longer limited by the requirement of practical utility, and that the range of acceptable answers is no longer limited by the requirement of consistency with prevailing doctrine. However, within a scientific context, new requirements are imposed in place of the old ones. Empirical inquiry becomes subject to the requirement that the questions it seeks to answer have theoretical relevance, i.e., relevance to the formulation or evaluation of scientific systems. Knowledge becomes subject to the requirement that the answers it provides be testable.

In pre-scientific contexts, questions may be rejected because they are "useless" and answers may be rejected because they are "heretical." In science, these restrictions are removed, and new ones imposed: questions may be scientifically rejected because they have no theoretical relevance, and answers may be rejected because they are not testable or because they fail to "pass" the appropriate tests.

The old requirements were characteristically unlimited in their potential range of application. Useless questions and heretical answers were rejected generally, not merely for certain specialized purposes. The new requirements differ strikingly from the old ones in this respect: they are recognized as pertaining only to the specialized context of science. Within the context of scientific inquiry, questions have no place unless they have some theoretical relevance, but scientists as well as other people generally recognize that questions which lack theoretical relevance may nevertheless be quite properly asked and answered, although outside the framework of pure science. Similarly, although non-testable answers to questions have no place within science, nevertheless many scientists are quite willing to recognize such answers as legitimate in contexts other than that of science, e.g., in religion. Science is thus a manifestly specialized activity, which does not necessarily displace pre-scientific patterns of thought, with the exception of those particular and limited aspects of pre-scientific thought which are incompatible with it.

The new requirements imposed upon knowledge systems within a scientific context differ from the traditional pre-scientific requirements not only in being more specialized, but in another important respect as well: these new requirements, unlike the older ones, can never be satisfied in a demonstrably permanent way, but rather suggest a developmental process for which a stable terminal point cannot be assumed. An answer to one question merely raises new questions, and may also cast doubt on answers already given to previous questions. Thus science has an intrinsically self-reinforcing effect in certain important respects.

The initial emergence of scientific systems, as described in preceding sections, was dependent on challenges from, and diversities within, the larger cultural environment. But the stimuli which played such a crucial part in the initial emergence of science, which were provided by the passage of accumulated knowledge through a sequence of different cultures, and by intra-societal diversities, have become less important as science has come to produce its own internal stimuli for its own development. Established scientific systems have come to be challenged not merely occasionally by historically accidental events in the larger cultural environment, but routinely on the basis of empirical evidence collected under conditions shaped by these systems themselves. Furthermore, the search for empirical evidence relevant to the testing of scientific systems leads to diversified inquiries in directions which practical technical interests might never have suggested.

The developmental process which is thus implied by the nature of scientific systems involves periodic transformations in which established scientific systems are displaced by new ones. Transition from one such system to another can take place in either of two sequences:

1) Discovery of an "anomaly" — i.e., an empirical fact inconsistent with predictions made on the basis of the prevailing system, or a defect in the logical organization of that system — may stimulate formulation of alternative systems.

2) Or, discovery that the facts accounted for by the established system can also be explained by an alternative system may stimulate acquisition of evidence which will provide a basis for selecting between the systems in question.

In either case, the outcome may sometimes be a re-affirmation of the established system, with apparent anomalies interpreted in a manner consistent with this system, and with proposed alternative systems rejected. Sometimes, on the other hand, the outcome is a rejection of the established system and its replacement by a new system, which then prevails until it, in turn, is itself displaced in a subsequent cycle of change.

It is well known that scientists have commonly resisted such transitions.[10] Scientists often ignore proposed alternatives to an established system unless persistent and serious anomalies suggest that something is importantly wrong with it. On the other hand, even very serious anomalies are not likely to lead to widespread rejection of an established system unless an acceptable alternative is available. Abandonment of an established system is likely to occur only when there are *both* serious anomalies *and* an acceptable alternative; one of these alone does not usually suffice. Furthermore, ambiguities in the criteria for identifying "serious" anomalies and "acceptable" alternatives provide considerable opportunity for scientists to justify, to themselves and their colleagues, what may appear retrospectively to be excessively rigid resistance to change. When a transition is successfully made from one system to another, all scientists in relevant areas of specialization do not necessarily participate in the change: sometimes, as Max Planck suggested, an established system is abandoned not because its supporters are converted to a new one, but because they "eventually die." [11]

We thus have two seemingly quite different possible consequences of the establishment of a scientific system. On the one hand, such a system may suggest new inquiries which are likely to lead, eventually, to the displacement of that system by another. On the other hand, the established system, by virtue of the achievements which led to its initial acceptance, and by virtue of vested interests in its persistence which may subsequently develop, is likely to retain the loyalty of a body of scientists who will respond conservatively to initiatives directed toward its displacement.

These two consequences are not mutually incompatible. In fact, efforts by scientists to bolster an established system may lead to discoveries which discredit that system instead. This paradoxical phenomenon is illustrated by the response to the Newtonian system during the two centuries of its dominance. It was commonly assumed during that time that the Newtonian system was a true and permanent foundation of physics, and it was in the context of efforts to apply that system more broadly that its defects unexpectedly emerged.

Of course, conservative support for an established system will not facilitate displacement of that system unless this support is manifested in research of the sort which will expose that system's weaknesses. If an established system is merely assumed to be "true" and not used as a basis for further research, the effect of support for the system will be to preserve its status as the supporters intended. For example, if sociocultural conditions in the post-Newtonian period had discouraged further scientific research, the Newtonian system could have constituted a final culmination of science.

Thus, some resistance to change is advantageous to scientific progress, but not all such resistance. In assessing any episode of resistance to scientific change in the past, we should take into account the limitations of available knowledge at the time the episode occurred. For example, the late nineteenth century resistance to the idea that the human species emerged through a process of evolution in the Darwinian sense appears quite reasonable insofar as such resistance was based on the fact that Darwinian evolution required a time span longer than the total age of the sun and the earth as calculated by some distinguished scientists of that day.[12] However, even when allowance is made for limitations of knowledge at various times in the past, much of the resistance which new and subsequently accepted ideas have initially confronted appears quite unwarranted in terms of scientific standards of evaluation. Thus, Darwinian evolution was challenged in the nineteenth century not only on the basis of reasonable scientific arguments as indicated above, but also on the basis of dogmatic theological assumptions which some scientists, along with many non-scientists, firmly adhered to.

It is possible that unwarranted resistance to new ideas in science may be declining, that with the maturation of science and the clarification of the distinction between scientifically relevant and scientifically irrelevant criteria of evaluation the rejection of new ideas within the scientific community may be less erratic and less prejudicial than it was in earlier periods in scientific history. In any case, we may note a tendency, among those who reflect philosophically on the nature of science, to allow for future conceptual transformations in a way in which they have not been allowed for in the past. It has often been implicitly assumed in the past that appropriate scientific concepts could be derived directly from observed facts. Einstein has attributed such a view to Newton and to pre-twentieth-century scientists generally.[13] By contrast, it has come to be recognized in the philosophy of science in the twentieth century that observational facts by themselves do not uniquely determine the concepts which may be appropriately applied to such facts, and that scientific concepts are thus permanently tentative in the sense that the possible desirability of abandoning any concept in favor of an alternative concept cannot be precluded.[14]

The development of this insight in the history of science and philosophy has been a protracted and complicated process. In the early seventeenth century Descartes explicitly presented his cosmology as a useful fiction rather than as a proclaimed truth. This anticipation of the twentieth century point of view was apparently only a device for avoiding persecution by the Church, and did not reflect Descartes' own view.[15] In more recent times, the spectacular success of science has stimulated religious organizations to re-examine their doctrines, to abandon ideas not deemed religiously essential which are contradicted by scientific theories, and generally to relinquish claims to authoritative knowledge concerning matters which appear to belong within the proper jurisdiction of science. In earlier days, as illustrated by the problem of Descartes and by his reaction to it, the situation was precisely the reverse: it was scientists who had to be careful to avoid stepping improperly across the boundary line between the domains of science and of religion. The fact that scientists had to be cautious in this way, compelled them to be "modest" in formulating the meanings and implications of their discoveries and theories.

Later, when pressure from the Church was no longer a major problem, scientists commonly became more extravagant in their claims, as illustrated by Hermann von Helmholtz (1821-1894) who assumed that physical science in the nineteenth century was approaching completion of its mission.[16] Reversion to the earlier scientific "modesty" of Descartes was brought about not by renewed pressure from any church or other structure external to science, but rather by a dramatic event within the framework of science itself: the displacement, in the early twentieth century, of the Newtonian system which had dominated science for over two hundred years and had come to be widely regarded as a permanent fixture. The shock of this event has encouraged twentieth century philosophers and methodologists of science to emphasize that all scientific concepts and laws have an intrinsically tentative status.

Most scientists are actually much less ready to question or to abandon their favorite systems, than such philosophical considerations might suggest. Nevertheless, the idea that they should be ready to do so on appropriate occasions has been strongly built into the value system of twentieth century science. It thus appears reasonable to suggest that all conceptual transformations in scientific history may not be structurally the same; that one such transformation may prepare scientists for others in the future and thus reduce the impact of others when they come; that the process of transition from one scientific system to another is thus a process which itself undergoes transitions over time.

NOTES

1. The Aristotelian system approached a "scientific" status in that it involved abstract formulations based on observation, but the integration between rational and empirical aspects was grossly imperfect in two respects. On the one hand, the observations utilized as an empirical basis for this system, were observations under ordinary conditions rather than controlled conditions, e.g., the observation that a heavy object will generally fall faster than a light one, under ordinary atmospheric conditions (as distinct from controlled conditions approaching a perfect vacuum). On the other hand, the abstract formulations of the Aristotelian system were drastically confined by dogmatic presuppositions not recognized as testable, e.g., the assumption that the heavens are perfect and that celestial objects engage only in perfect motions. The Newtonian system was

not only superior to the Aristotelian in predictive capacity; it also involved a more adequate integration between empirical and rational aspects, with controlled observation linked to formulations which had become largely freed from the requirement of conformity to prevailing cultural d⋯⋯ ⋯⋯ ⋯r, by twentieth-century standards, the Newtonian sys⋯⋯ ⋯ only outmoded science, but also in certain respects incompletely scientific: for example, Newton's assumption that concepts such as "time" and "space" were intuitively clear and did not need to be defined, obscured an ambiguity in the meaning of "simultaneity" which ultimately proved to be a decisively important defect.

2. Such a definition is suggested by Egon Bittner, "Radicalism and the Organization of Radical Movements," *American Sociological Review*, v. 28, no. 6, December 1963, pp. 928-940.

3. Distinctive features of Western culture are discussed by William H. McNeill, *The Rise of the West*, Chicago: University of Chicago Press, 1963.

4. For a discussion of the role of selective cultural diffusion in early scientific development, see George Devereux, "Psychoanalysis as Anthropological Field Work: Data and Implications," *Transactions of the New York Academy of Sciences*, Ser. II, v. 19, March 1957, pp. 457-472.

5. Karl Mannheim, *Ideology and Utopia*, trans. by Louis Wirth and Edward Shils, New York: Harcourt Brace, 1936.

6. Florian Znaniecki, *The Social Role of the Man of Knowledge*, New York: Columbia University Press, 1940.

7. For descriptions of ancient Babylonian astronomy, see O. Neugebauer, *The Exact Sciences in Antiquity*, New York: Harper 1962; B. L. van der Waerden, "Basic Ideas and Methods of Babylonian and Greek Astronomy," in A. C. Crombie, ed., *Scientific Change*, New York: Basic Books, 1963, pp. 42-60; and Derek J. de Solla Price, *Science Since Babylon*, New Haven: Yale University Press, 1961, chapter 1, "The Peculiarities of a Scientific Civilization," pp. 1-22.

8. This effect of totalitarianism is mentioned, with specific reference to the Soviet Union, by Bernard Barber, *Science and the Social Order*, Glencoe, Ill.: Free Press, 1952, pp. 119-120.

9. This convergence between the two traditions is discussed by

Stephen F. Mason, *A History of the Sciences*, New York: Collier, 1962.

10. This phenomenon is discussed and illustrated in detail by Bernard Barber, "Resistance by Scientists to Scientific Discovery," Science, v. 134, no. 3476, September 1, 1961.

11. These and other aspects of resistance to scientific change are discussed by Thomas S. Kuhn, *The Structure of Scientific Revolutions*, Chicago: University of Chicago Press, 1962.

12. For a detailed discussion of resistance to Darwinian evolution, see Loren Eiseley, *Darwin's Century*, Garden City, N. Y.: Doubleday, 1958.

13. Albert Einstein, *Ideas and Opinions*, New York: Crown Publishers, 1954, pp. 273-274.

14. In the long run, the course of scientific development is primarily shaped by observed features of nature. However, in any given situation, there is always more than one possible theoretical interpretation of observed facts, and hence always an aspect of scientific systematization which is arbitrary as far as observations of nature are concerned. It is important to recognize that there is no contradiction between these two statements.

15. See A. C. Crombie, *Medieval and Early Modern Science*, v. II, Garden City, N.Y.: Doubleday, 1959, pp. 217-218.

16. Albert Einstein and Leopold Infeld, *The Evolution of Physics*, New York: Simon and Schuster, 1961.

Chapter 5

THE CHANGING RELATIONSHIP BETWEEN SCIENCE AND SOCIETY

Requirements for the Persistence of Science

The origin of science involved special cultural conditions which have been discussed in the preceding chapter. The persistence of science, after its origin, does not necessarily require the same conditions. It does, however, require certain other conditions, primarily organizational rather than cultural, in the society in which the scientific process is to take place.

The basic conditions which science needs for its continued functioning may be summed up in two words: freedom and support. Freedom, in the form which is relevant here, is a negative condition, involving the absence of restraints which would interfere with scientific activity, and particularly the absence of restrictions on communication of scientific ideas and criticisms. Support, on the other hand, is a positive condition, involving the availability of resources for scientific activity, e.g., money for salaries, equipment, and supplies, and also esteem for scientific achievement, which facilitates recruitment of scientists.

The problem of freedom for science is as old as science itself. In recent times it has become an especially critical problem in societies with strong scientific traditions which have acquired totalitarian regimes, including especially Germany under Nazi rule and the Soviet Union in the Stalinist period.[1] It remains today a major problem in countries with authoritarian political systems, including the Soviet Union which, as the only scientifically-advanced nation in the world with relatively long and continuing experience under comparatively stable authoritarian rule, provides a unique setting for studies of the effect of limitations of freedom on scientific progress.

In the open societies of the West, the problem of freedom for science does not appear in its severest forms, although it has been an important problem in some contexts.[2] It persists, mildly but pervasively, particularly within the context of the bureaucratic organizations within which much scientific research is done. In certain respects, the ethos of science is similar to that of bureaucracy. For example, the idea that a properly-conducted scientific experiment should yield essentially the same results regardless of the personal style of the experimenter parallels the bureaucratic idea that organizational functioning should not be substantially affected by personality traits of individual bureaucratic officials. There is a corresponding similarity between scientific and bureaucratic emphases on standardized quantitative measurement. Such similarities facilitate the assimilation of scientific research into bureaucratic settings, which potentially threatens those aspects of scientific activity which require freedom from bureaucratic control, e.g., aspects such as selection of research topics, interpretation of findings, and communication of ideas and research findings within the scientific community.

Freedom is not something which is ordinarily given to scientists specifically. Rather, it is a condition which is ordinarily available to scientists only because it is available more generally. Furthermore, freedom is commonly not "given" to people at all, but rather emerges and persists as a by-product of organizational conditions involving a division of power among competing forces.[3] This has an important implication here, for if freedom is not something which society gives specifically to scientists, then we do not need to search for an exchange pattern, i.e., to ask what scientists give to society in return for freedom. By contrast, *support* for science is more specifically allocated, and hence may by presumed to involve an exchange pattern. It is therefore appropriate, in discussing the support which science receives, to ask why such support is given. This is a difficult question to answer. The difficulty arises from the fact that scientists in their most characteristically scientific activities, i.e., in pure scientific research, do not generally produce results which are predictable in advance, useful from a practical standpoint within a short time, and particularly useful to those who support the research. This state of affairs would hardly be attractive to sponsors.

In the past, much scientific research has required only very little support. However, modern large-scale science requires considerable external support. To a limited extent, the familiar model involving services exchanged for support is applicable in a straightforward way: i.e., there are some situations in which pure scientific research is supported because it can be expected to yield practically useful results within a reasonably short period of time, under conditions which will be distinctively beneficial to sponsors. This, however, is not the typical case, and to account for the flourishing of science in an era when science is expensive, we must allow for other mechanisms of support, including the following:

1) Similarity between the requirements of applied research and the requirements of pure scientific research sometimes makes it possible for scientists who receive funds for applied research to use those funds in ways which will serve pure-science research interests also, at relatively little extra cost. In other words, the cost of pure scientific research can sometimes be drastically reduced, and even sometimes eliminated altogether, by combining it with applied research for which support has been made available. Furthermore, the lack of complete clarity in the distinction between pure scientific research and applied research may make it possible, under some conditions, for supporting funds to be diverted from the latter to the former, in ways which may not be completely consistent with the expectations of those who are supporting the research.

2) Scientists characteristically want opportunities to engage in pure scientific research. Therefore, to some extent, sponsors of

applied research may offer such opportunities to their scientist-employees, in partial compensation for their services in applied research. Employers or sponsors may be able to provide such opportunities at relatively low cost, if the facilities which are needed have already been acquired for applied-research purposes.

3) "Benefits" from scientific research may come, in some contexts, to be defined broadly enough to include more than technological applications, and with such a broader definition, pure scientific research may acquire a degree of manifest relevance which it would not otherwise have. For example, the people of a nation may value the national prestige which comes from having major scientific discoveries attributed to one of their fellow citizens, and anticipation of such prestige may provide a basis for national support of scientific research. While the practical benefits of a major scientific discovery may be uncertain, long delayed and so widely diffused that sponsors of the research are not distinctly rewarded for their sponsorship, the prestige benefits are likely to be more predictable, more prompt, and more clearly localized.

Massive societal support for science is a relatively new phenomenon. Therefore we cannot yet know whether such support will be maintained over a long period of time on a stable basis, and, if so, what the mechanisms, conditions and rationale of such support will be. However, since science is able to utilize support greatly in excess of that which it is likely to receive from even the most affluent and generous societies, the maintenance of support for science is likely to continue to be a problem.

It should be understood that freedom and support will not suffice to produce scientific advances for which the necessary cultural background of ideas does not yet exist. Freedom and support are merely organizational conditions which are relevant in determining the extent to which science makes advances for which cultural preparation is adequate. Furthermore, the relationship between these conditions and scientific progress may be more complicated than is sometimes allowed for. Some freedom and some support are essential if science is to flourish. Nevertheless, important scientific achievements have occurred under circumstances which would appear exceedingly disadvantageous in terms of freedom and support. We cannot even be confident that maximal levels of freedom and support would be optimal for scientific progress. Fear of persecution apparently encouraged some early modern scientists to present their ideas not as absolute truths but merely as useful hypotheses, in the form which has since come to be recognized as most appropriate for science, and which is now characteristic of scientific presentations even in the complete absence of persecution. It has been suggested that psychology emerged in nineteenth century Germany because physiologists, unable to find satisfactory employment in their own field, moved into philosophy, and applied the concepts and techniques of physiology to philosophical problems.[4] These examples illustrate the possibility that conditions which are difficult for scientists might sometimes be beneficial to scientific progress. However, despite such complexities, the general importance of freedom and support as conditions facilitating persistence of scientific progress is beyond dispute.[5]

The Growth and Impact of Science

In the long history of human culture, there have been a number of relatively brief and localized episodes of intense cultural creativity, separated by lengthier intervals of relative cultural stability or "stagnation." Perhaps modern science will eventually appear, in retrospect, to fit into this pattern. However, from the perspective available to us at the present time, modern science appears to diverge strikingly from the familiar pattern of spasmodic but unsustained cultural creativity. The emergence of modern science in the seventeenth century stimulated a creative development which, despite temporary and local variations, has continued with great vigor for more than three centuries and has assumed massive proportions.

This growth of modern science has involved 1) continued internal progress within initially established scientific disciplines, such as physics, 2) a diffusion of scientific premises, methods, and concepts to new disciplines (e.g., the emergence of the social sciences in recent decades), 3) a diffusion of science from the cultural context of its origin in western Europe, to various other societies and cultures around the world,[6] and 4) spectacular increases in numbers of scientists, numbers of scientific publications, and expenditures for scientific research.[7]

The growth of modern science has been part of a broader cultural trend in Western society, which has produced, in aspects of Western culture other than science, features comparable to those found within a scientific context. For example, modern machinery is characteristically constructed so that a part can be replaced by a spare part with no significant change in functioning. Modern bureaucracies are characteristically organized so that particular incumbents of offices are similarly "standardized" in their official behavior, and hence replaceable without organizational change. One of the ideals of scientific organization is quite similar: a scientist who performs an experiment and gets certain results should ideally be replaceable by any other competent observer, without any effect on the observed results. We have here a general emphasis, in modern Western culture, on standardization of parts, which applies equally when we refer to parts of a machine, to officials as parts of bureaucratic organizations, and to scientific observers as parts of the social system of science. We have, in other words, an example of a feature of modern scientific organization which is also, more broadly, a feature of the Western culture in which science matured and in which it remains largely embedded.

The preceding discussion of the emergence of the scientific process has emphasized the highly specialized character of science as an aspect of the culture of society as a whole. Looked at from one point of view, such specialization makes science increasingly different from other aspects of culture. On the other hand, the specialization of science may also be considered as merely one among a number of parallel specialization processes within the broader pattern of Western cultural development. Looking back on proto-scientific formulations of the pre-modern era, we inevitably notice what appears, from our modern vantage point, to be confusion (i.e., imperfect differentiation) between scientifically revelant and scientifically irrelevant features. For example, we find Copernicus (1474-1543) describing the sun as occupying a "royal throne" around which the planets circle as his "children." We find, similarly, in pre-modern times and in relatively unmodernized societies today, comparable "confusions" pertaining to other areas of culture: e.g., judicial systems in which the aim of clarifying points of abstract law is not clearly separated from the aim of reconciling opposing parties in disputes; markets in which personal ties and traditional customs interfere heavily with strictly economic aspects of exchange; and patterns of government in which the roles of secular and ecclesiastical authorities are not clearly separated. Modern Western culture has encouraged differentiation tendencies in these and other institutional areas. Thus, the tendency for science to purify itself through a sharpened distinction between scientifically relevant and scientifically irrelevant considerations is part of a much larger movement which also involves comparable processes in aspects of culture far removed from science.

The fact that science is thus part of a larger cultural trend makes it quite difficult to distinguish genuine effects or impacts of science on society, from other phenomena which may be neither effects of science nor causes of it, but which are instead associated with science as parallel aspects of the same larger pattern of which science is also an aspect. However, two kinds of effects of science can be distinguished with reasonable clarity: the impact of science on traditional knowledge, and its technological impact.

1) Modern scientific knowledge is actually more compatible

with traditional knowledge, from a strictly logical standpoint, than is commonly assumed. Thus, as Max Gluckman has pointed out,[8] African witchcraft beliefs concerning the cause of disease are largely compatible with scientific explanations: one can accept the modern scientist's idea that typhus is transmitted by lice, while also assuming that the lice are sent to selected victims by practitioners of witchcraft. Similarly, modern scientific knowledge appears reasonably compatible with prevailing religious beliefs in the West which are still recognizably related to religions of pre-scientific times, e.g., with liberal modern interpretations of Catholic and Protestant theologies. Nevertheless, although the points of direct incompatibility between scientific and non-scientific ideas have been much more limited than is usually recognized, the conflicts produced by such incompatibilities have been profoundly important, and have affected traditional beliefs much more fundamentally than would have been necessary merely to reconcile them in a purely logical sense with scientific knowledge.

2) Science has made possible immensely potent new technologies, which have created critical new problems of social organization and control. The general character of these problems may best be illustrated by considering the potential effects of a kind of technology which is still (in 1971) in a very early stage of development, and to which society has therefore not yet needed to make an adjustment. Let us suppose that it becomes technically possible to control the weather within broad limits on a seasonal and regional basis, e.g., to provide the northeastern United States with a severe winter or a mild one, and a wet or dry summer. Given such a new technical capacity, problems of the following sorts would be likely to emerge:

a) Different people would want different kinds of weather, and diverging interests in this respect would presumably create new political cleavages.

b) Long-range interests of society might differ from dominant short-range preferences. Thus, most people might want "sunny and mild" weather every day, but this could be disastrous in the long run. Thus, issues of societal discipline vs. immediate gratification would be superimposed upon issues arising from the clash between opposing groups with incompatible meteorological interests.

c) Expansion of the capacity to control natural events often outstrips the expansion of knowledge concerning long-range effects of such control. Thus, after we have learned to control the weather, we might still not know what the long-range effects of any changes which we initiate will be. Perhaps, for example, hurricanes, tornadoes, and other meteorological phenomena which men generally would be inclined to dispense with, may have important functions of which no one is aware.

d) Boundaries of political jurisdictions commonly fail to coincide with boundaries of natural phenomena which may become subject to artificial control. Perhaps, for example, weather in the United States could not be controlled without thereby also influencing weather in other countries.

Such problems may be summarized in the statement that new technologies tend to impose upon the decision-making apparatus of society new burdens which this apparatus is poorly prepared to handle.

Prior to the twentieth century, societal reaction against science focused primarily on the apparently subversive character of new scientific ideas, on the potentially destructive effect of science on traditional knowledge and values. The crises produced by such anti-scientific reaction have largely abated, not merely because science has "won" its battles on this front, but also because the victory of science has turned out not to entail necessary defeats for certain important values which antedate science. Thus, the major religions of the West, after a protracted period of confusion, have accommodated themselves to modern science with reasonable success. At the present time, societal reaction against science has come to focus more on the destructive impact of the new technologies which science has made possible. The continued flourishing of science may well depend on the ability of science to dissociate itself in popular thinking from manifestly destructive applications of scientific knowledge, and to offer more hopeful prospects to those upon whose goodwill it depends, just as science in the past has dissociated itself from the assumption that the knowledge it produces is generally incompatible with values which have come down to us from pre-scientific ages.

Possible Futures of Science

No scientific system can ever be proven to be "true" in any absolute sense, or to be scientifically superior to all possible rival systems. The strongest claim that can properly be made for such a system is that it is superior to those particular rival systems which have been proposed thus far, and that it is not confronted by major anomalies. Thus, even if it happens that a given scientific system remains accepted for a long period of time, and no reason for abandoning it has been found, there is nevertheless no basis for classifying that system as "permanently established," and denying the possibility that some day its abandonment will appear appropriate.

Science is thus potentially an endless process as far as its own internal structural features are concerned. This does not mean, however, that science will actually continue endlessly. It could come to an end in any or several ways: through the destruction of the civilization of which it is a part; through a collapse of the arrangements which provide essential freedom and support; or through the apparent, or actual, exhaustion of its potentialities, i.e., through the absence of unsolved basic problems remaining within the domain of pure science.

The last two of these possible causes of the termination of science could be combined. Thus, a scientific system might emerge which is truly impressive and appears eminently satisfactory, which has defects that could be uncovered through sustained inquiry, but which remains firmly entrenched because conditions of freedom and support do not permit such inquiry. If such a system should become permanently fixed as a final triumph and culmination of science, we could attribute the end of science partly to the qualities of this system and partly to the absense of social conditions which would permit further probing to determine the limitations of these qualities. In fact, if the course of history had taken a turn slightly different from that which it actually followed, the Newtonian system might have brought an end to science in this way, as some distinguished scientists assumed that it would. It is not at all difficult to imagine a society in which Newtonian physics is accepted as a "final" product of physical science, and utilized for various practical purposes, but without additional scientific inquiries of the sort that might test the Newtonian system and that actually led to its displacement in the early years of the twentieth century.

The fact that "completions" of science which have been anticipated in previous centuries have failed to materialize, does not constitute a reasonable basis for denying that similar predictions made now for the future might turn out to be correct. Presumably, however, if science in its most basic areas should achieve a stable synthesis, there will remain numerous problems of technical and historical interest, and numerous opportunities for discovery through exploration, which may provide stimuli for continued research in a broad sense, even if this research is not classifiable as pure science.[9]

If science on a large scale does continue into the distant future, without achieving a "final" formulation, and also without encountering social conditions which would make its further progress impossible, there are several possible ways in which this future science could be related to the larger society, including the following:

1) Science could continue to maintain an uneasy, unstable relationship with society, and continue to have a disruptive impact on major social institutions.

2) Science could become clearly "institutionalized" as a subsystem of society, integrated with other subsystems on a stable basis, performing specialized functions for society as a whole and receiving societal support in return. This would require that society be reorganized in a way which would permit it to adapt routinely, without disruption, to changes which the continued progress of science might stimulate in the future.

We have already achieved a reorganization which permits us to adapt, more readily than before, to strictly intellectual implications of new scientific developments. In the early seventeenth century, a critic rejected one of Galileo's discoveries on the ground that the week had been divided into seven days, named after the seven planets, and that if more planets were added, "this whole system falls to the ground."[10] Today, discovery of a new planet would not, in itself, produce comparable intellectual disruption in our society. We have come to make allowance, in advance, for the possibility of such discoveries, so that when they occur they can be accepted routinely. However, we have not yet developed institutions capable of adapting without disruption to the technological implications of science.[11]

3) Science could cease to be merely a specialized activity with which only a small segment of society is directly concerned, and could become instead a major concern of society as a whole, with "science for its own sake" ceasing to be merely the ideology of a single profession, and becoming transformed into a major societal goal.

Modern science required, for its initial emergence, a special set of cultural conditions. Once modern science was established, it acquired a self-reinforcing tendency which has produced massive scientific growth during more than three centuries. However, science has also been producing effects which cannot yet be clearly classified in terms of their ultimate implications for science itself. We must thus allow for several possible futures, and especially for the possibility that science may turn out to be either self-reinforcing or self-terminating in the long run.

NOTES

1. For descriptions of relevant episodes pertaining to the Soviet Union under Stalin, see Conway Zirkle, *Death of a Science in Russia*, Philadelphia: Univ. of Pennsylvania Press, 1949; Zhores A. Medvedev, *The Rise and Fall of T. D. Lysenko* (translated by I. Michael Lerner), New York: Columbia University Press, 1969; David Joravsky, *The Lysenko Affair*, Cambridge, Mass.: Harvard University Press, 1970; and John V. Murra, Robert M. Hankin, and Fred Holling, *The Soviet Linguistic Controversy*, New York: Kings Crown Press, 1951.

2. See, for example, Edward Shils, *The Torment of Secrecy*, New York: Free Press, 1956, and C. P. Snow, *Science and Government*, Cambridge, Mass.: Harvard University Press, 1961.

3. For a general discussion of the social conditions under which freedom is most likely to be maintained, see Gerard DeGre, "Freedom and Social Structure," *American Sociological Review*, October, 1946, pp. 529-536.

4. Joseph Ben-David and Randall Collins, "Social Factors in the Origins of a New Science: This Case of Psychology," *American Sociological Review*, August, 1966, v. 31, n. 4, pp. 451-465.

5. Many different social conditions have affected the progress of particular sciences in particular times and places. For a detailed discussion of this topic, see Joseph Ben-David, *The Scientist's Role in Society*, Englewood Cliffs, New Jersey: Prentice-Hall, 1971. I have mentioned here only what appear to be the most generally relevant conditions, and only those which pertain primarily to the persistence of science rather than to its origin.

6. This process has been analyzed by George Basalla, "The Spread of Western Science," *Science*, v. 156, May 5, 1967, pp. 611-622.

7. See discussions of the quantitative growth of science in Derek J. de Solla Price, *Science Since Babylon*, New Haven: Yale University Press, 1961, and in the same author's *Little Science, Big Science*, New York: Columbia University Press, 1963.

8. Max Gluckman, *Custom and Conflict in Africa*, New York: Free Press, 1955.

9. The possibility that science might achieve a stable culmination has been recently discussed by Bentley Glass, "Science: Endless

Horizons or Golden Age?," *Science*, January 8, 1971, v. 171, no. 3966, pp. 23-29.

10. Francesco Sizi, quoted by Gerald Holton and H. D. Roller, *Foundations of Modern Physical Science*, Reading, Mass.: Addison-Wesley, 1958.

11. For discussion of possible adaptations of this sort, and of several possible futures for science, see John Rader Platt, *The Step to Man*, New York: Wiley, 1966.

Chapter 6

THE ORGANIZATION OF THE SCIENTIFIC COMMUNITY

The Concept of "Scientific Community"

What we call the *scientific community* consists of all the scientists of the world, collectively, insofar as they maintain among themselves patterned relationships organized to facilitate the scientific process.

The scientific community is internally differentiated into disciplines and areas of specialization within disciplines, and also into national groupings. For some purposes, it will be convenient to use the term scientific community to refer to what is actually a segment of the total world scientific community, consisting of the scientists in a given country and/or discipline.

However, the scientific community is not defined simply in terms of its membership. Scientists may be organized collectively for a variety of purposes. An association of scientists could become the equivalent of a labor union, functioning primarily to protect the economic interests of its members, or it could become a political pressure group, seeking changes in government policies. In totalitarian societies, scientists may be organized by political authorities for the purpose of maintaining centralized control over activities of scientists. Even if a structure devoted to one of these purposes should have a membership consisting of all scientists in a given country, we would not have an example of what is meant by the scientific community here. The scientific community in the present sense exists only to the extent that scientists are organized to facilitate the scientific process itself, in relatively direct ways.

This aspect of our definition requires us to recognize that a scientific community exists as a relatively concrete and autonomous structure only in some societies and at some historical periods. In pre-modern times, the lack of any clear distinction between the role of "scientist" and various related roles such as "philosopher" and "technologist" was paralleled by a lack of any patterned arrangement for scientifically relevant communication among those who participated in activities which we today would differentiate as "scientific." In some totalitarian societies of the twentieth century, associations of scientists have been so powerfully integrated with ruling elites that they cannot be considered as constituting what are defined here as scientific communities; instead they are structures in which the characteristic functions of the scientific community are quite imperfectly differentiated from, and firmly subordinated to, other functions which belong outside the scope of the scientific community.

In the scientifically advanced open societies of the West, the scientific community has attained its greatest degree of autonomy. Even so, we should not expect to find anywhere in the real world (as distinct from the world of abstract models) structures which are characterized by perfect differentiations of function. Even under conditions which are maximally favorable to its autonomous development, the characteristic functions of the scientific community may be imperfectly differentiated from other functions.

Our attention here will be directed primarily to the functioning of the scientific community in its relatively "pure" forms. Clarification of the ideal features and operations of the scientific community will provide a standard in terms of which various existing structures of social organization and control among scientists can be analyzed and evaluated.

The characteristic functions of the scientific community may be defined initially on a residual basis. Certain requirements of the scientific process, such as financial support, are generally satisfied by structures outside the scientific community, i.e., by employers or sponsors. Scientific research is undertaken by individual scientists and research teams. There remain, however, functions which cannot be effectively performed either by structures external to science, or by scientists organized only as individuals and research teams, but which must rather be performed by scientists collectively, through arrangements from which there is no reason for any competent scientist to be excluded. These, by definition, are the central functions of the scientific community.

As a basis for identifying these functions, we may note certain features and implications of the exchange system which links scientists to external structures such as employers and sponsors of scientific research. In this system, scientists provide such structures with information, and receive, in return, facilities for further research, and rewards including financial remuneration.

As long as science is dependent upon external structures for support, an exchange system of this general type is indispensable for the continued flourishing of science. However, such an exchange system, by itself, would tend to degenerate into an arrangement clearly detrimental to science. On the one hand, information obtained by isolated scientists and research teams, and communicated directly and exclusively to external structures, would lose much of its potential scientific value, both because it would not receive the scrutiny of competent and disinterested scientists other than those involved in the immediate exchange itself, and because it would not be integrated with information obtained by other scientists elsewhere. On the other hand, external structures tend to distribute rewards in exchange for information received, not on the basis of an evaluation of this information in strictly scientific terms, but on the basis of diverse other criteria, including apparent practical usefulness. If the rewards coming from external structures were the only rewards available to motivate scientists, their motivations would presumably tend to shift away from pure science and toward practical applications of a highly rewarded sort.

The scientific community functions to preserve the integrity of science which is thus challenged by certain implications of the exchange system linking scientists with external structures. It does this by maintaining an exchange system of its own, linking scientists to each other, which may be called the *scientific exchange system*.

In this exchange system, the scientific community receives, evaluates and integrates scientific information from its numerous participants in different specializations, thus providing a body of scientific information which is much more coherent, extensive and scientifically relevant than the information that would otherwise be available. The scientific community also distributes among its participants rewards such as professional recognition which are based primarily on achievements of a strictly scientific nature, thus protecting science from what might otherwise be the almost complete dominance of reward systems based on non-scientific criteria such as immediate practical usefulness of information.

The Scientific Community and External Structures

As a basis for explaining the functions of the scientific community, I briefly mentioned a hypothetical situation in which the scientific community did not exist, and in which there was accordingly no exchange system among scientists to supplement the exchange system between scientists and external structures. It is possible also to imagine a situation at the opposite extreme, in which the scientific community as a unit obtains significant resources from the society, which it is free to allocate internally among its participants in terms of exclusively scientific criteria. This extreme situation, like its opposite, does not exist in reality. Scientists may be deeply involved in making decisions about the allocation of resources among different specializations and different research projects, but they characteristically do so within the context of expectations that such decisions will reflect the practical interests of society, or of some segment of society, in addition to, or in place of, interests of a purely scientific sort.[1]

The situation which actually prevails, in open societies with strong scientific traditions, is roughly intermediate between the two extremes outlined above. A scientific community exists with a reasonable degree of autonomy, and performs its characteristic functions, but does not have at its disposal significant resources which it is free to allocate exclusively in terms of the criterion of the advancement of pure science. In fact, the model which most clearly reveals the characteristic situation of the scientific community in societies such as our own is a model which assumes that resources which support science come exclusively from external structures, and that the scientific community functions within the limits imposed by this condition.

A scientist is likely, then, to be involved in exchange systems of two types: one which relates him to employers and sponsors, and another which relates him to the scientific community. There is, of course, considerable variety among the situations of different investigators in this respect. At one extreme are some who relate exclusively to employers; these would have a doubtful status as "scientists." At the other extreme are scientists involved exclusively in the exchange system which links them to scientific colleagues, with no special, separate obligations to employers or sponsors. Among those who do participate in two separate exchange systems as described here, some find that they can do so with relatively little conflict, while others are exposed to sharply conflicting expectations.[2]

The scientific community is relatively intangible compared with the external structures with which scientists interact. Such structures are likely to be relatively concrete units such as business corporations or government agencies. The scientific community, by contrast, appears as an abstraction which, at best, may be partially approximated in the real world under comparatively favorable conditions. Furthermore, the money and facilities which scientists receive from employers and sponsors are certainly more tangible than the professional recognition which they receive from the scientific community. The lesser tangibility of this exchange system internal to the scientific community makes it difficult to focus attention on that system as distinct from the one which links science to external structures. Nevertheless the scientific exchange system is more centrally related to the scientific process and has a greater theoretical significance for the sociology of science.

The Scientific Exchange System

Exchange systems have been identified in diverse areas of social life.[3] An analysis of the scientific exchange system may therefore provide an opportunity to bring the analysis of the organization of the scientific community within the same general framework which is applicable also in other sociological analyses.

In a simple bilateral exchange, there are two parties, A and B, and two commodities or services, one of which is given by A to B while the other is given by B to A.

The parties which enter into the scientific exchange system may be identified as follows:

1) The term *scientist* may be taken to include anyone who participates directly in the scientific process by making relevant contributions. For some purposes, research teams may be classified as equivalent to individual scientists, as far as their participation in the exchange system of science is concerned.

2) The *scientific community* includes contributors to science generally. Ordinarily a scientist's primary relationship is not with the entire scientific community but with a narrowly specialized segment of it. However, this distinction, although crucial for some purposes, will not be particularly relevant here.

The scientific exchange system involves exchanges of benefits as follows:

1) Relevant contributions by a scientist to the scientific community must pertain in some direct way to the advancement of the scientific process. Presumably, such contributions must be informational in a broad sense. They may consist of theoretical innovations, empirical discoveries, methodological improvements, or critical evaluations of contributions of other scientists. Of course there is wide variation in the importance of different

contributions to science. At one extreme, there are a few contributions which have become major landmarks in scientific history, e.g., the achievements of Galileo, Newton, Darwin, Mendel and Einstein. At the other extreme are a very large number of minor contributions which attract the attention of small numbers of specialists only briefly.

2) One might say that ideally participation in scientific activity should be intrinsically satisfying, to the extent that additional rewards are not needed as incentives. In reality, however, additional rewards play an important part. An effective reward for scientific contributions must have several characteristics. a) It must, of course, be considered valuable by the scientists who are intended to be its recipients, and for whom it is intended as an incentive. b) On the other hand, as long as there are no "licensing requirements" which limit the number of people who may attempt to participate in science, an effective reward should be relatively valueless to most other people who lack scientific competence, and whose attempts to "participate" in the scientific exchange system on a large scale would overwhelm the scientific communications system and thus impede scientific progress. c) It should be a kind of reward which the scientific community has at its disposal to allocate among its members. d) It should be obtainable only, or only on a large scale, from the scientific community, in order that the community may maintain control over the distribution of scientific rewards.

Money, as a type of reward, satisfies the first of these criteria, in that it is generally valued by scientists. However, money fails to satisfy the other criteria: it is valued not only by scientists but by people generally; it is not available to the scientific community to distribute among scientists on a significant scale; and it is more readily available from sources other than the scientific community. Esteem, in the most general sense, is not much more adequate than money in terms of these criteria. However, one particular form of esteem, professional recognition, satisfies all the specified criteria excellently. It is highly valued by scientists; it is generally meaningless and valueless to people outside the relevant professional (scientific) community; it is readily available to the scientific community to allocate among its members; and it is not available from sources outside that community. Professional recognition may not be the only type of reward which satisfies the relevant criteria and which is actually employed, but it is one such reward, and a major one. For present purposes we may assume that it is the only one, apart from the intrinsic satisfaction of participation in the scientific process, even if this is an oversimplification of the actual situation. Of course, professional recognition may lead to additional rewards of other types, including money. However, such additional rewards may generally best be recognized as coming primarily from sources other than the scientific community, and hence as not part of the exchange system of that community.

The identities of the parties which enter into the scientific exchange system, and of the benefits which are exchanged in this system, have certain implications concerning the way in which the exchange system operates, as follows:

1) In some exchange systems, the terms of exchange are agreed upon in advance, sometimes with arrangements for "fixed prices," and sometimes through bargaining. In other systems, characteristic of gift-giving rather than of the market, goods or services are exchanged without explicit attention to their respective values, and one party gives to the other without any explicit commitment to do so and without any explicit expectation concerning a gift to be received in return. The scientific exchange system generally fits into the "gift-giving" type.[4] Scientists are presumed to do their research and to share the resulting information with colleagues, not in expectation of specific rewards, but because they want to do so regardless of reward. Furthermore, informational contributions on the one hand and professional recognition on the other are notoriously difficult to evaluate in standardized quantitative terms, e.g., in terms of money, and this is quite characteristic of ideal "gifts." On the other hand, contri-

butions to science are subject to rigorous critical evaluation of a sort which is not usual for "gifts," even though this evaluation is not on a standardized quantitative basis which would permit precise ranking of scientists in terms of the values of their respective contributions.

2) There is another way in which the exchange system of science may be classified. Some exchanges are structurally symmetrical, as when A and B give each other gifts with no significant distinction between types of gifts exchanged and no patterning of the sequence in which the gifts are given. The scientific exchange system is asymmetrical in both respects. Its exchanges are characteristically of the "performance-reward" type. That is, such exchanges involve performances, on the one hand, for which rewards are given in return. In these exchanges, there is a typical sequence: performances, which must meet certain standards in order to be acceptable, come first, and the rewards come afterward.

The Normative System of Science

The exchange system in the scientific community is regulated by standards or expectations concerning the behavior of participants. Such expectations are technically known to sociologists as *social norms*. Several sociologists have enumerated what they consider to be the important social norms of the scientific community.[5] The conception of scientific activity as organized around an exchange process provides a basis for a more systematic analysis of these norms.

A bilateral exchange between A and B entails four processes: 1) A gives something to B, 2) B receives and accepts what A gives; 3) B gives something to A; and 4) A receives and accepts what B gives. In the exchange system with which we are here concerned, the four processes are as follows: 1) scientists, individually and in groups (research teams), contribute relevant information to the scientific community, or the appropriate specialized portion of that community; 2) the scientific community receives and accepts relevant contributions; 3) the scientific community distributes professional recognition among its members; and 4) scientists receive and accept professional recognition as allocated by the scientific community.

Each of these four processes is subject to normative controls of at least two sorts. First, each of them is controlled by norms which pertain to the identification of the parties entering into the exchange system. Secondly, each is controlled by norms pertaining to the benefits to be exchanged.

The norms pertaining to identification of relevant parties may be summarized as follows:

1) The scientist is expected to make his contributions to the scientific community directly, giving the scientific community priority over other potentially or actually interested parties. This means, specifically, no secret hoarding of information, no distribution of relevant information to employers, sponsors or others until after distribution within the scientific community, and no general publicity pertaining to such information until after it has been communicated through appropriate scientific channels.

2) The scientific community is expected to receive and accept relevant information from any source, without irrelevant distinctions in terms of such criteria as the contributor's nationality, ethnic background, professional status, or personality. This norm does not preclude a strong element of scepticism or suspicion when a purported discovery is announced by someone of doubtful competence, but ideally such negative attitudes should not preclude consideration of the contributed information on its merits. The norm does mean that, in principle, anyone may enter into the scientific exchange process and hence into membership in the scientific community.

3) The scientific community is expected to allocate its rewards equitably among those who have made relevant contributions.

4) The scientist is expected to recognize the scientific community as the only appropriate source of judgment concerning scientific achievement and the distribution of rewards for such

achievement. This does not mean that a scientist is expected to agree invariably with his colleagues, concerning the merits of his own or of another's contributions. Nor does it rule out intelligent discussion of scientific issues in a wider public which includes non-scientists. It does mean that a scientist is expected to recognize that any error in evaluation of a scientific contribution is to be corrected within the scientific community itself, and that approval or disapproval of a "scientific" idea by the general public or by political authorities is not scientifically relevant.

The norms pertaining to the benefits to be exchanged may be summarized as follows:

1) The scientist is expected to transmit scientifically relevant information, as distinct from information selected on the basis of technological relevance, conformity to prevailing dogma, or personal prejudice, and he is expected especially to refrain from presenting information of a fraudulent nature.

2) The scientific community is expected to evaluate received contributions on the basis of scientific criteria, as distinct from criteria pertaining to technological relevance or conformity to dogma.

3) The scientific community is expected to distribute only rewards considered appropriate, such as professional recognition. This means, in particular, that a scientist's reward for achievement should not be of such a nature as to give him power or privilege of a sort which might subsequently interefere with the participation of his actual or potential rivals in the scientific exchange system.

4) The scientist is expected to refrain from seeking inappropriate types of rewards from the scientific community; for example, he should avoid utilizing professional eminence as a basis for obtaining privileges which would interfere with the appropriate participation of others in the scientific exchange system.

I have not attempted here to offer anything approaching an exhaustive enumeration of norms commonly accepted within the scientific community. Rather, attention has been directed here only to those norms which are clearly related to certain basic specified features of the scientific exchange system. These norms appear to have a central importance which may justify the claim that they constitute a core around which the normative system of the scientific community is organized.

Some Complexities in the Scientific Normative System

It is well known that actual behavior often violates normative standards, or in other words that there is often inconsistency between norms and behavior. It is also well known that there are often inconsistencies within a normative system itself, e.g., that norms pertaining to appropriate behavior in particular situations may be incompatible with more general norms or values. Furthermore, it is well known that there may be certain complexities within a normative system which do not entail any inconsistency: e.g., norms requiring different behavior from different people, or from the same person in different situations. It is less commonly understood, however, that many normative systems include a particular form of "complexity without inconsistency," which involves not only an expectation of different behavior in different situations, but also a different evaluation of such situations themselves. Thus, behavior which is considered normatively ideal under actual existing, non-ideal conditions may differ from behavior considered ideal under ideal conditions which do not exist. For example, in one very familiar normative system, "military disarmament" represents an ideal national policy under ideal conditions of secure peace, while "military preparedness" represents an ideal policy under actual, non-ideal conditions of external military threat.

The distinction between ideal behavior (norms) pertaining to ideal conditions, and ideal behavior (norms) pertaining to actual conditions, is important because, unless we allow for it, we are likely to misinterpret both prevailing norms and observed behavior. Specifically, in the absence of this distinction, we would be likely to interpret observed behavior as normatively "deviant" when it is inconsistent with norms pertaining to ideal conditions, even though it may be fully consistent with norms pertaining to actual conditions. We would also be likely to assume erroneously that stated norms are nothing but hypocritical verbalizations simply because they are generally violated, unless we recognize that such norms may pertain only to ideal conditions and hence may not be relevant in the interpretation of observed behavior in actual situations.

The distinction between norms pertaining to ideal conditions, and norms pertaining to actual conditions, may be applied in analyzing the benefits exchanged in the scientific exchange system, i.e., on the one hand, scientific performances or achievements, and on the other hand professional recognition as a reward. A case could be made for the proposition that under conditions more ideal than we can actually expect, certain other elements might more appropriately enter the exchange system, and that the norms which actually operate concerning the above-mentioned benefits pertain to actual conditions deviant from ideal ones:

1) Scientists receive professional recognition from the scientific community primarily as a reward for scientific success rather than for conformity to scientific norms. Since non-normative variables such as "luck" are among the determinants of success, emphasis on success rather than conformity as the primary basis of reward means that a discrepancy characteristically exists between the normative system and the reward system of science. Thus, a successful scientist who is careless about adhering to some of the peripheral norms of the scientific community is much more highly rewarded than one who adheres scrupulously to all relevant norms but never succeeds in making a significant discovery — or who does make discoveries but finds that someone else has made them and announced his results first.

Presumably it would be better from the standpoint of scientific progress to have rewards given for conformity to relevant norms than for success, since success is a function of luck as well as excellence, and rewards for luck are functionally superfluous. On the other hand, it is better to have rewards given for scientific success than for conformity to inappropriate norms. In fact, a heavy emphasis on normative conformity as a basis for reward within the scientific community would be likely to push the actual norms of that community farther away from what they should be in the interest of scientific progress by encouraging ritualistic elaboration of such norms in ways which are scientifically irrelevant but which would facilitate invidious distinctions in terms of degree of conformity. By contrast, although problems may arise concerning ways of measuring success in contributing to science, the emphasis on success as a basis for reward encourages continuous critical examination of institutionalized norms in terms of their relevance to reward-producing achievements, thus preventing these norms from diverging too far from what they should be if scientific activity is to be performed with maximal effectiveness. A certain limited discrepancy between criteria for evaluating performance under ideal conditions and criteria for evaluating it under actual conditions is thus apparently functional for science.

2) A scientist may seek or receive any of three kinds of reward for his work: a) rewards inherent in scientific activity, e.g., the satisfaction of contributing to science, b) rewards not inherent in scientific activity but distributed within and by the scientific community, e.g., professional recognition by other scientists, and c) rewards obtained primarily from extra-scientific sources, e.g., money and public recognition. Ideally, one might say, the scientist should seek primarily not to be given rewards by others but simply to contribute to science because he finds this to be intrinsically rewarding. However, such motivation by itself may not suffice to produce scientific contributions. Given this non-ideal motivational condition, it may be appropriate that the norm pertaining to ideal conditions be relaxed to permit the scientist to seek the second kind of reward as well as the first.

Robert Merton has described the ambivalence of scientists

with respect to questions of priority and professional recognition. He interprets this ambivalence as a reflection of an inconsistency between two values, one of which leads a scientist to want his discoveries to be properly credited to himself, while the other leads him to modestly admit how little he has accomplished.[6] Norman Storer interprets this ambivalence differently, pointing out that professional recognition, although strongly desired by scientists, may interfere with competent response to one's work by other scientists, which is also strongly desired.[7] Warren Hagstrom offers a third interpretation, pointing out that scientific contributions are gifts and that compensation for gifts is characteristically not explicitly demanded.[8] We have here a basis for another explanation of such ambivalence. Seeking recognition from other scientists on the basis of the priority of one's own work is "better," in terms of the normative system of science, than seeking money or recognition from the general public, but also "not as good as" working merely for the satisfaction of contributing to science. The fact that a kind of reward which is highly valued by many scientists thus stands intermediate between two other kinds of reward in terms of acceptability within the scientific normative system may account for at least some of the ambivalence which Merton has documented.

The relations among these four interpretations may be summarized as follows. Storer assumes that the scientist will be ambivalent about *receiving* professional recognition. The others assume only that he will be ambivalent about *actively seeking* such recognition. According to Merton, the scientist hesitates to seek recognition because a norm of modesty calls upon him to minimize his accomplishments. According to Hagstrom and the present interpretation, the source of ambivalence is a normative requirement for minimizing the need for extrinsic rewards rather than minimizing the importance of the accomplishments for which rewards are given. The present interpretation also includes the idea that, to the extent that professional recognition is sought, this is not merely because such recognition is pleasurable but also because it is normatively more acceptable than certain alternative kinds of reward, even though less acceptable than certain other alternatives.

The distinction between norms pertaining to ideal conditions, and norms pertaining to actual conditions, has implications concerning not only the benefits exchanged in the scientific exchange system, but also concerning the identities of the parties involved in relevant exchanges. Thus, under certain non-ideal conditions, it may be normatively appropriate for scientists to communicate primarily with outside agencies rather than with the scientific community, and appropriate also for certain scientists to receive rewards disproportionate to their contributions.

1) The secrecy imposed on research associated with the development of the atomic bomb during the Second World War might have been reasonably defended as consistent with the appropriate norms of science under the conditions then existing. Confronted with the fact that eminent scientists strongly committed to the norms of science engaged in such secret research on behalf of the United States and allied governments, we do not need to assume either violation of scientific norms or inconsistency within the normative system. Instead, we may interpret such behavior as a normatively appropriate response to a non-ideal situation. The norm prohibiting secrecy was not "violated," according to this interpretation, but rather was temporarily suspended, while nevertheless retaining its status as the appropriate norm for ideal, as distinct from actually prevailing conditions.

2) Merton has brought to the attention of sociologists the so-called Matthew effect, first noted, apparently, in the Gospel according to Matthew.[9] This effect is, essentially, a tendency for rewards to be given to those who have already been well-rewarded, i.e., to those who need the rewards least: "For unto everyone that hath shall be given, and he shall have abundance, but from him that hath not shall be taken away even that which he hath." In science, the Matthew effect means that a contribution made by an already prominent scientist will receive more

attention and recognition than a comparable contribution by a more obscure scientist. Presumably, under ideal conditions, it would be "best" for science if all contributions received attention according to their apparent merits, and if rewards were distributed accordingly. However, the flow of scientific information is so voluminous that scientists have difficulty in keeping up with developments in their respective fields. This makes it impossible for all contributions to receive the attention they "deserve." Under these conditions, as Merton has suggested, it may be most efficient to have special attention directed to contributions by men already known to be exceptionally competent and productive, even if this is "unfair" to some other scientists and to their contributions.

Rapidly changing conditions in science and society are now producing changes in the normative system of the scientific community. One source of change is the widespread recognition among scientists and others that "misuse" of scientific knowledge can have overwhelmingly disastrous consequences. This situation has led some scientists to conclude that they should do more in some respects, and less in other respects, than the norms of science as traditionally interpreted might suggest: i.e., that they should take action as scientists to influence the application of scientific knowledge, and also that they should avoid engaging in, and discourage, scientific research which seems likely to lead to primarily "undesirable" technological applications. It is not possible to predict with reasonable confidence what the ultimate outcome of this and other current movements will be. However, the concepts and distinctions presented above may perhaps be helpful in analyzing whatever changes in the normative system of science ultimately emerge.

Some Organizational Implications

We come now to consider ways in which the exchange system of science, together with features of the scientific process itself, may help us to understand various features of the social organization of the scientific community.

We may consider, first, the delineation of the boundary of the scientific community, and then its internal organization. The boundary of the scientific community involves both a distinction between "members" and "non-members," and a distinction between scientifically relevant activities and other activities of members. The internal organization of the scientific community encompasses both patterns of relationships among members, and patterns of differentiation among them.

I have pointed out that the exchange system of science is of the "performance-reward" type, with scientists' performances consisting of the acquisition and presentation of relevant information, and the rewards consisting of professional recognition. The nature of the required performance limits the range of potential membership in the scientific community, and the nature of the ensuing reward limits the extent to which members can participate in the activities which characterize the community. These limitations may be summarized as follows:

1) Membership in the scientific community, or, in other words, participation in the scientific exchange system, is limited to those who are able to make relevant contributions. There are, in principle, no other restrictions; in particular, no restrictions in terms of such manifestly irrelevant criteria as nationality or cultural background. Only a relatively small percentage of the total population of even the most scientifically advanced societies is abe to meet the qualifications for participation or membership. Those who do meet these qualifications, are rather widely scattered around the world (with some concentration in Europe and North America), and represent a diversity of national and ethnic backgrounds.

2) Professional recognition has been identified here as the typical reward which the scientific community allocates among its members in return for their contributions. But no matter how

pleasurable professional recognition may be to those who receive it, this particular reward can hardly be considered sufficient; scientists must also have financial support and facilities. The fact that these characteristically come from sources outside the scientific community, e.g., from employers or sponsors, means that the participation of a scientist in the scientific community and its associated exchange system is usually intertwined with his participation in other, closely related but quite different structures. A scientist may, for example, be employed by a business corporation or a governmental agency, have his salary and facilities provided by his employer in exchange for *technologically*-relevant (as distinct from scientifically-relevant) information which he provides; may nevertheless extract the scientifically-relevant components of the information he is thus paid to provide to his employer, and present these components to the scientific community, receiving professional recognition in return. As this hypothetical example suggests, participation in the scientific exchange system is often peripheral to a scientist's occupational role.

The requirements of acceptable scientific performance thus give the scientific community a widely scattered membership, while the intangibility of the rewards distributed by that community force its members to seek support elsewhere and thus commonly gives their participation in the community an occupationally peripheral or "part-time" character. In these respects, the scientific community has features diametrically opposed to those of traditional communities, which are geographically and culturally localized and which encompass most of the social involvements of their members.[10]

Within the scientific community, relationships among scientists are both *cooperative* and *competitive*. The importance of cooperation is reflected in the large percentage of scientific research reports which have more than one author, and the large percentage in which the assistance of scientists other than the author(s) is acknowledged. The prevalence of competition is reflected in the long history of disputes over priority of discovery,[11] and in documented instances of "races" between competing research teams.[12]

Cooperation among scientists is necessary for the organization of scientific activity, for the production of relevant contributions which scientists make to the scientific community. On the other hand, the distribution of rewards creates a situation which tends to elicit competition. We thus find that one side of our exchange process implies cooperation while the other implies competition.

Just as each side of the scientific exchange process has a different implication with respect to *relations* among scientists, so also each side has a different implication with respect to *differentiations* among scientists. In the scientific exchange system, contributions from scientists consist exclusively of innovations, in the sense that no scientist may "contribute" information already contributed by someone else, and the rewards which scientists receive in return consist in principle of nothing more than esteem. This means, on the one hand, an exceedingly complex pattern of specialization in scientific activity, i.e., in the "production" of relevant contributions, and on the other hand an exceedingly simple "stratification" system resulting from the distribution of esteem as a reward.

No scientist in modern times is *merely* a "scientist"; every one is highly specialized in terms of various distinctions among scientific disciplines and sub-disciplines. In addition to such disciplinary distinctions — e.g., "physicist," "biologist," "sociologist," and their respective sub-categories — there are other role distinctions based on specialization in terms of skill rather than subject matter, e.g., the distinction between theoretical physicists and experimental physicists. However, such skill-specializations are comparatively weakly developed, while specialization on a subject-matter basis is one of the most prominent features of the organization of scientific activity.

I have pointed out that every contribution to science, if it is to be accepted as a genuine contribution, must be original and hence unique in some relevant respect. This requirement means that each scientist's activities must be differentiated in a unique way, in addition to ways which are established in roles. The unique contributions of each scientist are, in effect, individualized refinements of role-specializations of a primarily "disciplinary" sort. The difference between a specialized scientific role and one of its individualized manifestations may be illustrated as follows. If we say that Albert Einstein was a "physicist," we are identifying a specialized disciplinary role which he occupied within the scientific community. If we say that he formulated the theory of relativity, we are thereby identifying not a "role" but a unique personal contribution which represents an individual specialization within a role.

The diverse disciplinary specializations of science are interrelated in such a way that developments in one may have implications for others. For example, in ancient times the idea that the rising and setting of sun and stars is caused by rotation of the earth was proposed but effectively challenged on the ground that such rotation would produce violent winds and cause people to fall away from the earth, which manifestly did not happen. Acceptance of a rotating earth entailed a radically new physics, which was not adequately developed until the seventeenth century.[13] Similarly, although Darwin's theory of evolution emerged largely from his personal biological observations, the theory had to be evaluated in terms of its astronomical and geological implications, as well as its biological implications; the length of time which this theory required for the evolution of man was much greater than the maximum possible lifetime of the sun and the maximum possible age of the earth as calculated by some leading nineteenth century authorities on these topics.[14]

Given the exceedingly high degree of disciplinary specialization in science, together with the interrelations among developments in different specialized areas, we might seem to have, in the case of science, an outstanding example of "division of labor." Paradoxically, the opposite is true: several sociologists have pointed out that there is relatively little division of labor in science.[15] To understand this, we must keep in mind two distinctions mentioned earlier, which cut across each other: the distinction between specializations on a "subject-matter" or disciplinary basis, and on a "skill" basis; and the distinction between specialized roles and unique individual specializations *within* the various roles. The concept of division of labor pertains to relationships among differentiated roles; hence unique individual specializations are not relevant in assessing the applicability of that concept. Role specializations on the basis of skill, as in the above-mentioned case of theoretical and experimental physicists, do involve division of labor but are comparatively unimportant in science as a whole. Role specializations on a disciplinary basis are overwhelmingly prominent in science, but these do not involve systematic patterns of interdependence such as "division of labor" entails: thus, the combination "physicist-biologist" cannot be classed with such combinations as buyer-seller, leader-follower, husband-wife, or physician-patient. *Particular* developments in one scientific specialization may have particular implications for another specialized area, as illustrated above, but this does not mean a systematic dependence of one of the relevant specialized roles on the other.

In striking contrast to the complexity of specialization in scientific activity, but in line with the simplicity of the scientific division of labor, the reward system of science is in principle quite simple. There is only one major type of reward involved, namely professional recognition, which is a form of esteem rather than prestige, i.e., it applies to particular individual people rather than to roles. There is also essentially only one recognized criterion for the distribution of this reward, namely contribution to science. We thus have a "stratification" system which involves little more than differences in esteem on the basis of achievement.[16]

The simplicity of the scientific stratification system, and the simplicity of the scientific "division of labor" which is somewhat obscured by a complex pattern of disciplinary specialization, are both associated with other kinds of structural simplicity. The scientific community has, in principle, no formal structure, no structuring on a kinship basis, and no geographical or territorial boundaries. Characteristic exchanges among its members involve, on the one hand, uniquely individualized contributions, and, in return, intangible rewards consisting mostly of esteem. In these respects the scientific community resembles a primary group — i.e., an informal group with intimate relations among its members. On the other hand, its strong task-orientation, the high objective standards of performance which it maintains in the evaluation of potential contributions by its members, and the geographical dispersion and diversity of cultural background of its members, suggest an organizational pattern radically different from that of ordinary primary groups. This combination of attributes suggests a "professional" pattern, but the scientific community differs from that pattern also, as I have previously explained, by virtue of its major emphasis on exchanges among its own members rather than exchanges with external clients, and by virtue of its peripheral relationship to the occupational role system of society.

The scientific community, if it is to perform its functions effectively, must be organized in a way which is consistent with the requirements of the scientific process, and consistent also with the kind of relationship which science must have with the larger society in order to maintain sufficient support. The uniqueness of the patterns which have emerged in response to this challenge is a reflection of the unique nature of the challenge itself.

NOTES

1. For an illustration of a way in which scientific and nonscientific criteria may be applied jointly in allocating financial support

among different scientific disciplines, see Alvin M. Weinberg, *Reflections on Big Science*, Cambridge, Mass.: M.I.T. Press, 1967.

2. For an analysis of different ways in which scientists may be related to the scientific community on the one hand and to external structures on the other, see Roger G. Krohn, *The Social Shaping of Science*, Westport, Conn.: Greenwood Publishing Co., 1971.

3. See, for example, George C. Homans, "Social Behavior as Exchange," *American Journal of Sociology*, v. 62, May 1958, pp. 597-606; Alvin W. Gouldner, "The Norm of Reciprocity: A Preliminary Statement," *American Sociological Review*, v. 25, April 1960, 161-178; and Peter M. Blau, *Exchange and Power in Social Life*, New York: Wiley, 1964.

4. This aspect of the exchange system of the scientific community is analyzed by Warren O. Hagstrom, *The Scientific Community*, New York: Basic Books, 1965.

5. The classic description of these norms which has served as a starting point for subsequent analyses is that of Robert K. Merton, *Social Theory and Social Structure*, Glencoe, Ill.: Free Press, 1957, chapt. 16, "Science and Democratic Social Structure," pp. 550-561. For a more recent formulation, see Norman W. Storer, *The Social System of Science*, New York: Holt, Rinehart, and Winston, 1966.

6. Robert K. Merton, "The Ambivalence of Scientists," *Bulletin of the Johns Hopkins Hospital*, v. 112, February, 1963, pp. 77-97.

7. Storer, *op. cit.*

8. Hagstrom, *op. cit.*

9. Robert K. Merton, "The Matthew Effect in Science," *Science*, v. 159, no. 3810, January 5, 1968, pp. 56-63.

10. For an explanation of the use of the term "community" to refer to structures lacking a geographic locus, see William J. Goode, "Community Within a Community: the Professions," *American Sociological Review*, v. 22, April, 1957, pp. 194-200.

11. This topic is discussed by Robert K. Merton, "Priorities in Scientific Discovery," *American Sociological Review*, v. 22, December, 1957, pp. 635-659.

12. See, for example, James D. Watson, *The Double Helix*, New York: Atheneum Press, 1968.

13. For an excellent description of the way in which different

ideas in ancient physics and astronomy were inter-related, see Herbert Butterfield, *The Origins of Modern Science*, New York: Macmillan, 1960.

14. See Loren Eiseley, *Darwin's Century*, Garden City, New York: Doubleday, 1958.

15. Hagstrom, *op. cit.*, Storer, *op. cit.*, and especially, Kenneth J. Downey, "The Scientific Community: Organic or Mechanical?," *Sociological Quarterly*, v. 10, no. 4, Fall, 1969, pp. 438-454.

16. There are prestige differentials among disciplines, which provide a basis for distinguishing between the "prestige" of a given scientist's disciplinary role and the "esteem" which he has acquired through his personal contributions. However, prestige differentials are relatively unimportant, compared to the distinctions which prevail in terms of esteem, and are largely derived from the latter: those disciplines in which major advances are being made, and in which scientists are thus acquiring considerable esteem, tend to be relatively high in prestige.

CONCLUSION

Science as described in preceding chapters possesses numerous combinations of features which appear paradoxical. It involves an empirical emphasis on close observation and accurate description of nature, but also a quite different rational emphasis on conceptualizations which go far beyond the range of observed facts. It is a cumulative process in the sense that it moves consistently in the direction of increased knowledge of nature, yet it is also non-cumulative or discontinuous in the sense that new scientific concepts are not simply "added to" older ones but often displace the latter. It involves a unique subordination of men to nature, in the sense that scientists ask questions of nature and commit themselves in advance to accept whatever answers they ultimately receive, yet it has also made possible an immense increase in the capacity of men to control nature. It involves an emphasis on creativity, in the sense that each scientific contribution is a unique addition to the culture of the scientific community, yet it also involves a contrasting emphasis on standardization, in the sense that any two scientists performing the "same" experiment are expected to obtain the "same" results. The formulations which it produces are public and are subject, in principle, to validation by any competent observer, yet the specialized knowledge and skill necessary to acquire relevant competence gives these formulations an apparent esoteric quality which contrasts sharply with their public and objective status.

It involves a very complex pattern of specialized activity, but a very simple division of labor. Its organization is inherently cooperative, in the sense that scientists "pool" their contributions, and must do so if science is to progress, yet it has a reward system which strongly encourages scientists to compete with each other for priority in making and announcing discoveries. Science is very old, with notable achievements in ancient times which can be distinguished retrospectively as "scientific," and with more than three centuries of continued development in its modern form, yet it is also exceedingly new as a heavily-supported, large-scale enterprise. Its cultural context is distinctively Western, yet it has a universal relevance which transcends cultural boundaries.

In addition, two kinds of contrasts pertaining to the concept of science may be singled out for special mention:

1) Science tends to have features quite different from those which science itself has identified in nature. Thus, science enables us to predict natural events, yet science itself is remarkably unpredictable as far as its own future development is concerned. Its formulations are characterized by great clarity and precision, yet the organizational arrangements of the scientific community are quite vague and imprecise by comparison with many other kinds of social organization which sociologists study.

2) When science is viewed macroscopically, with a long time span and a diversity of social contexts taken into account, it appears to have features different from those which emerge from comparatively microscopic perspectives. In the short run, we find social values and personal prejudices influencing the formulations of scientists, while in the long run the direction of scientific development is determined by aspects of the natural world. From a comparatively microscopic standpoint, science appears as a "method," while a more macroscopic view enables us to see it as a process which involves a method but which also has aspects not reducible to methodological rules. From a microscopic standpoint, science is an international or trans-national activity, which contrasts sharply with other, non-scientific elements with-

in each national society or culture. From a more macroscopic standpoint, the international aspect of science is overshadowed by the fact that it is heavily localized in a single basic civilization, that of the West, and is international primarily within that civilization alone; and the differences between science and other aspects of society are overshadowed by similarities which are distinctively Western, e.g., by the similar emphasis on quantitative standardization in Western bureaucracy and in Western science.

Such contrasts between different characteristics of science make it quite difficult to analyze science without appearing to make "contradictory" statements. These contrasts also make it difficult to describe science in a way which is appropriately balanced, without emphasizing one aspect while inadvertently overlooking another "opposite" aspect.

The development of an adequate sociological conception of science is complicated also by the narrow cultural and temporal distribution of large-scale science, which reduces opportunities for comparative observations; by the status of sociology as a branch of science, which encourages sociologists to view science from the perspective of the participating scientist himself rather than that of the sociological observer; and by one feature of science, in particular, which marks it as unique among social phenomena: the fact that its formulations are subject, in principle, to determination on the basis of observed aspects of the natural world.

This feature has a crucial implication: it means that science cannot, by definition, have its course of development determined by society. Knowledge which has its content primarily shaped by social forces rather than by observed features of nature is simply not science as the term is commonly used and as it is defined here. In the long run, society can encourage science or inhibit it, but not shape its course, or even predict its course on a long-range basis with reasonable confidence. On the other hand, while science thus stands apart from society as far as long-range social influence on its formulations is concerned, the intellectual and technological implications of science have stimulated drastic and disruptive transformations of society.

Its dynamic, disruptive character and its unstable relationship to society make it inappropriate to classify science as an institution which performs functions for society, or an occupation or profession whose members provide services to employers or clients, on a stable basis in exchange for support. Pure scientific research is commonly not a full-time gainful occupation, does not generally involve professional-type obligations to clients, and is not integrated with other aspects of society as traditional institutions are.

However, the fact that science when viewed macroscopically fits only awkwardly into sociological categories which entail a stable relationship between science and society does not preclude classification of science in terms of sociological concepts altogether. On the contrary, various kinds of sociological concepts are applicable to science, although in some cases only with special adjustments to allow for unique features of science as a social phenomena.

The concept of "cultural process" provides one way to bring science within a sociological frame of reference. Science, viewed as a cultural process, is a cognitive and a developmental process as well. Within this frame of reference, science may be compared with processes of cognitive development at the individual as distinct from the cultural level, with patterns of pre-scientific, traditional cultural knowledge which are generally stable rather than intrinsically developmental, and with forms of cultural development which involve primarily aspects of culture other than the cognitive aspect.

Modern science can also be recognized as similar, in several important ways, to other aspects of the Western civilization within which it emerged. Science is a highly specialized branch of culture, and in this respect it reflects a tendency in modern Western culture toward increased institutional specialization: a tendency which appears in political, economic, religious and

Conclusion

familial institutions as well as in science. Western culture is also noted not only for "radical" tendencies to follow the implications of accepted principles to whatever ultimate conclusions they may lead, but also for both persisting tensions and complex adjustments among differing radical tendencies. This pattern is reflected in the scientific integration between the differing radical traditions of rationalism and empiricism, and in the way in which the radical scientific emphasis on subordination of man to nature emerged in an historical context already characterized by a radical technological emphasis on domination of man over nature and a radical religious emphasis on subordination of man to God.[1]

It is true that science emerged in Western civilization as something importantly "new," yet it involved a combination of elements which were already quite old, and the break with the past which it entailed was itself consistent with a dominant tendency in Western culture which antedated science: a tendency toward contrast, instability, and sharp reversals of cultural emphasis.

Science may also be analyzed in terms of exchange systems in which it is involved, which are fundamentally similar to exchange systems found in other areas of social life. Science needs freedom from social control, and societal support; it offers to society, in return, knowledge which has tended to undermine cherished beliefs, and which often has no immediate useful application. This hardly appears to offer a promising basis for an exchange system between science and society, yet one has emerged nevertheless, involving devices by which societal interest in useful knowledge is "converted into" support for pure science. (For example, pure scientific research is combined with applied research, and definitions of "useful" knowledge are stretched to accommodate pure-science research interests.) This exchange system, which links scientists individually and collectively to the larger society, is reinforced by an internal exchange system within the scientific community, in which individual scientists "pool" their contributions, with each receiving, in return, the amount of professional recognition considered appropriate in terms of the value of his contribution. The two inter-related exchange systems have provided a basis for the recent flourishing of science, although this arrangement appears highly fragile and unstable, and is now subject to particularly great stress with the growing realization that in the absence of appropriate controls the technology which science has made possible constitutes an immense danger to mankind.

We thus have several ways in which science may be brought within a sociological frame of reference, and may be viewed as sharing important features with other phenomena with which sociologists are concerned. On the other hand, in each of these sociological perspectives, allowances must be made for aspects of science which are unique; in particular, for the special relationship between science and nature which limits the range of possible relationships between science and society.

NOTE

1. Lynn White, Jr., has suggested that Christianity, especially in its Western versions, encouraged a view of man as master of nature, and that this view has strongly influenced modern science and technology. ("The Historical Roots of our Ecologic Crisis," *Science*, v. 155, no. 3767, March 10, 1967, pp. 1203-1207.) I have proposed a different interpretation here: *technology* involves the dominance of man over nature but *science* involves a subordination of man to nature in the sense that scientists allow nature, as they observe it, to determine their formulations. This interpretation, in turn, is consistent with a conception of Western culture as incorporating and integrating sharply opposed tendencies.